MODERN ROOFING

CARE AND REPAIR

MODERN ROOFING

"Repair & Replace Flashing, Ventilation, Gutters, Downspouts, Dormers & Skylights"

PROJECTS

MODERN ROOFING

CARE AND REPAIR

DONALD L. MEYERS

CREATIVE HOMEOWNER PRESS®
A DIVISION OF FEDERAL MARKETING CORPORATION,
24 PARK WAY, UPPER SADDLE RIVER, NEW JERSEY 07458

Manufactured in United States of America

Current Printing (last digit)
10 9 8 7 6 5

Editor: Shirley M. Horowitz
Art Director: Léone Lewensohn
Illustrator: Norman Nuding

Cover photograph courtesy of Bird & Son

We wish to extend our thanks to the many designers, companies, and other contributors who allowed us to use their materials and gave us advice. Their names, addresses, and individual identifications of their contributions can be found on *page 144*.

LC: 81-66574
ISBN: 0-932944-34-5 (paper)
0-932944-33-7 (hardcover)

CREATIVE HOMEOWNER PRESS®
BOOK SERIES
A DIVISION OF FEDERAL
MARKETING CORPORATION
24 PARK WAY,
UPPER SADDLE RIVER, NJ 07458

FOREWORD

The first sign of trouble in the roof is a leak. If you are fortunate, seepage will be discovered before the leak becomes a damaging stream into the rooms below. Stains or dampness usually precede such a complete disaster, and can be detected by an annual inspection. A proper inspection tour will include both the exterior and the interior of the roof, as described in detail in Chapter 2. The more obvious signs of trouble are raised or broken shingles, a loss of mineral granules in the shingles, dark-looking spots in the sheathing, and other such symptoms discussed in Chapter 2.

When problems are detected, the obvious question is: What can be done? The easy—and expensive—answer is to call in a contractor and tell him to fix it. Assuming that the contractor is honest and competent, your troubles should be over, but your pocketbook may suffer from shock. In many cases, you have little choice but to call in a contractor. A steep or otherwise inaccessible roof leaves you no other option. If you are handicapped, there also is no alternative. Those who have flat, "built-up" roofs may be able to make minor repairs, but a brand new built-up roof requires special equipment and is not really in the province of the do-it-yourselfer.

However, most roofs can be repaired—and many even completely replaced—by the average homeowner. Although there are some cautions involved, reroofing is not a complicated job. It may be time-consuming, laborious, somewhat hazardous and boring, but it doesn't require anything more intellectual than the ability to measure, drive a nail, and be wary of falling off the edge. A strong back is a help too, but perhaps you can recruit a relative or friend for the muscle work. The good news is that, since roofing is a labor-intensive job, you can save about half of the costs by doing it yourself.

Furthermore, even if you have one of the aforementioned "impossible" roofs, there is no need to panic and call in a contractor at the first sign of trouble. The problem could well be a minor one which you can fix yourself. A leak near the edge of a roof is often caused by gutter defects, not the roof itself. A flat roof may need only a little patching up. Sometimes, all you need is a little more adhesive on the flap or a shingle or two.

But let's assume that, for one reason or another, you will need the services of a contractor. What type of contractor? You may well need a roofing contractor, but there are various types: residential or commercial, those who handle both, those who specialize in gutters and leaders (downspouts), and so on. If the problem is in the framing or sheathing, perhaps a general or carpentry contractor is best. And, if and when you call a contractor, how do you find a good one? That question is often unanswerable, but there are ways to avoid the obvious frauds. These precautions and considerations are discussed in Chapter 3.

What we will try to do in this book is to first identify the problem and put it in perspective. If necessary, details are given for emergency steps that can be taken in order to avoid a disaster. However, *Modern Roofing* also tells you how to conduct periodic inspections so that such emergency measures need never be adopted.

Chapter 2 enables you to detect trouble; Chapters 6 through 9 show how to fix the damage before it becomes worse. These sections even cover how to completely reroof a house or repair the eaves, as well as fix downspouts and gutters and add needed ventilation. Materials, their costs, availability, advantages and disadvantages are reviewed in Chapter 4—followed by tools and safety precautions in Chapter 5. Related topics, such as adding dormers, skylights and fans can be found in Chapter 11.

CONTENTS

1

ROOF DESIGN AND STRUCTURE

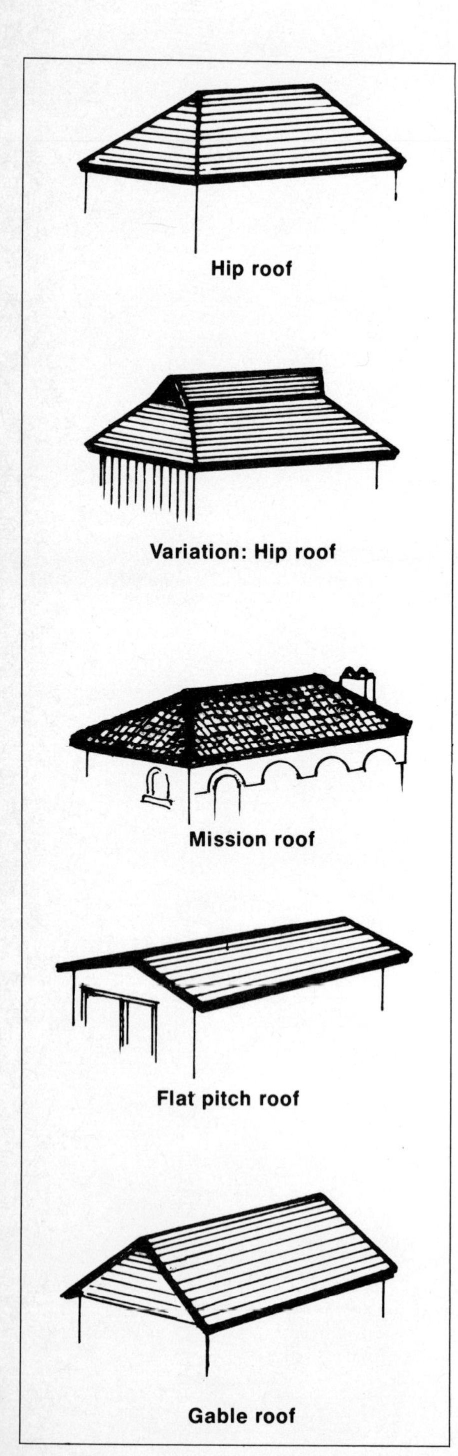

Roof design often is thought of as the arbitrary choice of the builder or developer. Unfortunately, this is all too often true, especially in large development tracts. One home will have a hip roof, another a gable, yet another a Mansard or gambrel thrown in for variety. In these cases, the design of the roof varies solely to make one house a little different from the next. Rather than having all the homes the same, with the chance of one owner stumbling into the bedroom next door, even this variety is probably better than nothing. But the roof design of a home should really be a product of the overall architecture.

Perhaps the most striking differences in home roofing result from geographical considerations. Houses in the American West or Southwest (and some areas in the South) need not be designed for snow loads or extensive cold spells, while these factors are a prime consideration in the North, most of the East, and the Midwest.

ROOF STYLES

Although there are variations of each, roofs generally follow one of the styles listed below.

Basic Designs

Flat. Many residences in the West and Southwest have flat roofs, because there is little snow. Flat roofs are made of built-up roofing, felt and hot asphalt, covered with a layer of stones or gravel.

Gable. This is the most common type of roof in temperate climates. The top of a gable roof divides the roof area into two sloping planes, designed to funnel snow and rain to the eaves, where it collects in gutters and is dispersed to the ground via downspouts, also known as leaders. The gables are triangular endwalls, with a ridgeboard between them and rafters (or roof framing) from the ridgeboard to the walls below, and with the ceiling joists at the bottom of the triangle. Inside the gabled area is the attic.

Working on a flat roof does not offer the dangers found in working on a steep roof. This type of roof cannot handle snow loads.

Hip. This is a roof with four sloping planes, eliminating the triangular endwalls at the front and back of a gable-roofed home. Since eaves are created on four sides, with rain and snow being diverted in all directions, gutters are required all around the roof perimeter. If the home were perfectly square, the roof would wind up in a point, as in a pyramid, and there would be no ridge board at the top. Since, however, a home with such a ''peak'' roof is very rarely built that way, there is ordinarily a small ridge board along the top.

Gambrel. Not common today, a gambrel roof is often found on ''Dutch colonial'' homes or in old barns. This is basically a gable roof where the slope is broken into two planes on each side. One

A roof like this one, with many planes and pitches, is hard to reroof and difficult to estimate accurately for materials.

A gable roof is the most frequently seen style in temperate climates. It directs snow and rain away from the house and foundation, and provides attic space inside.

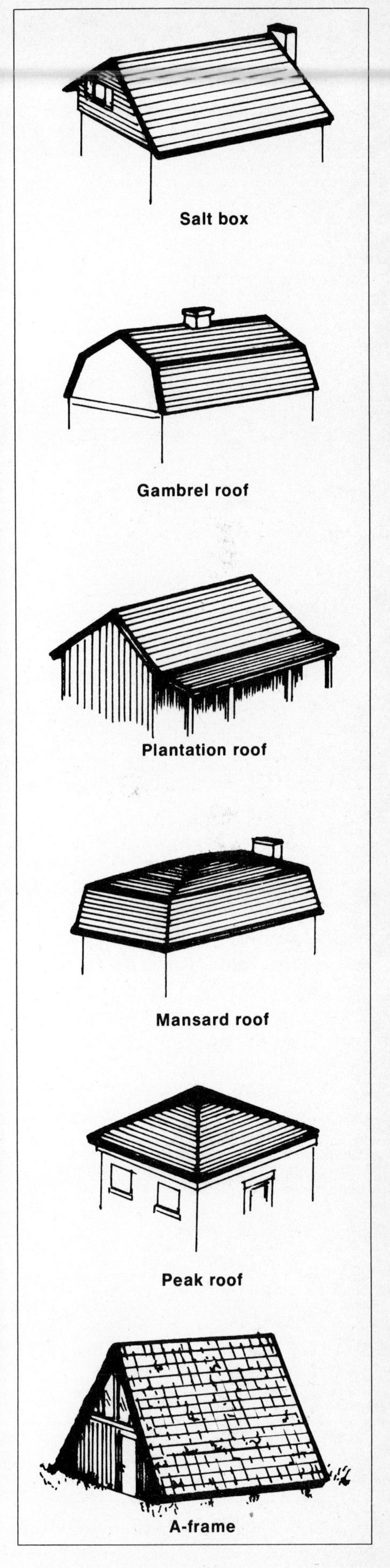

Salt box

Gambrel roof

Plantation roof

Mansard roof

Peak roof

A-frame

The three dormers on this gambrel provide additional interior living space and also emphasize the roof's attractive lines.

result is more and higher attic space. Usually, the planes on each side are broken approximately in the center, giving four separate roof areas of about the same size. The three angles at the top are approximately equal.

Mansard. Used in ''French provincial'' style homes, this roof is an exaggeration of the gambrel roof, resulting in a nearly flat roof in the upper planes, and a nearly vertical slope on the bottom planes. This leaves a great deal of roof in the ''attic'', which is generally used as another floor of living space. Mansard roofs almost always are accompanied by small gables along the steep side slopes. They frequently combine with a hip to form a rather complicated structure. The steep sides of Mansard roof are often treated more as part of the sidewalls than as part of the roof. This can result in difficult decorating and design decisions (such as whether to use roofing or siding materials), as well as complicated repair and reroofing problems for the do-it-yourselfer.

Low-slope roofs are sometimes composed of two or more shed-roof planes.

Shed. A simple roof sometimes used for outbuildings (such as a ''shed''). Since there is only one plane and one slope, a home would rarely be equipped with a single shed roof, although a combination of low-slope sheds are sometimes found. Shed roofs are popular for small additions. A double shed roof with a valley in between is called a ''butterfly''.

A-frame. This variation of the pitched roof is not really a roof, but a home with combined walls and roof. This design is used almost exclusively for vacation homes. The roof and the sidewalls of the home are the same, the main idea being to protect the home from heavy snows in ski country. The A-frame does sometimes show up in vacation homes in warmer climates. A reroofing job on an A-Frame isn't a minor undertaking, since the home is almost all roof, and is very steep.

Salt box. Many Early American homes were ''salt-box'' style, with two stories in front and one in back, creating an uneven roof. This style is occasionally copied today.

Dormers. A dormer gives the gable roof extra architectural elegance. It also enables the attic area (if high-pitched enough) to be used as living space. There are two types of dormers, shed and gable. A shed dormer extends over a large portion of the roof. Gable dormers are smaller, often only large enough for a window. There usually are two or more gable dormers together on each roof slope.

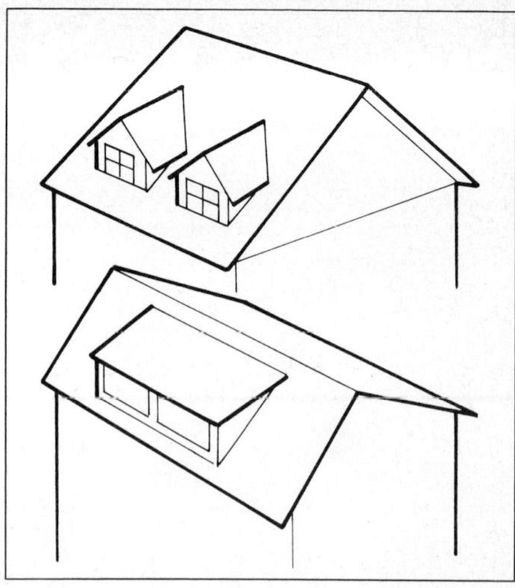

Gable dormers, smaller than shed dormers, usually are grouped on one roof plane.

Complicated Roofs

It is not so important to give your roof style a name as to determine how to fix or reroof it. Unusual designs may require a mathematician to estimate the amount of reroofing materials needed, or a roofing expert to figure out how to do it. A multiplaned house with turrets is not a do-it-yourself job.

STRUCTURAL COMPONENTS

No matter what the style, a properly constructed pitched roof consists of similar basic elements. There is a ridge board at the top (forming the peak), rafters between the ridge board and the outside walls, and joists running directly between the sidewalls.

Roof framing can be considerably more complicated than this, especially when dormers are involved, but basically the same elements are involved no matter what the style, and the same functions are performed. Any kind of framing is intended to provide strength and to give shape to the final product, as the bones do for the body.

Framing Requirements

It may seem that the joists are more a part

Salt box roofs, most commonly seen in New England, have one short plane that meets the second story and one long plane that extends down to the top of the first story.

of the floors below than part of the roof. In fact, they do serve as framing for the room ceilings below. But the joists also serve as a structural component of the roof itself. If you try to imagine the roof simply resting on the sidewalls, without joists and beams, you can easily see that—because of the slope of the rafters—all of the weight of the roof framing, decking and finishing materials is thrust down and out at the top of the sidewalls. Without the ceiling joists to securely fasten the walls together, this pressure would push the sidewalls out and down. The entire structure would collapse. This is one reason why sidewalls, joists, and rafters always must be firmly connected.

Collar beams also help in relieving some of the sideways thrust of the rafters. There should be a collar beam tying together every third set of rafters at a minimum. Even if the ceiling joists do not have much direct weight to carry from above, they must be sufficiently strong to prevent outer walls from spreading apart under the roof load.

For added strength, ceiling joists usually rest on, and are spliced together at, an interior load-bearing wall. Once all the outside walls and bearing walls are constructed and plumbed, then the ceiling joists are nailed in place. Only after the ceiling joists are firmly attached does the actual roof construction begin. Using a rather complicated technique, the angles of the rafters are calculated and precut to size. Each rafter is cut to fit exactly right at the ridge board and notched to fit over the sidewalls. The rafters are then nailed to ridgeboard, sidewalls and joists. Collar beams are nailed between each third set of rafters as mentioned previously. (If the slope is low, collar beams may be nailed between *each* set of rafters.)

Help for hips. Hip roofs require extra hip rafters at the corners. Jack rafters are then nailed between the hip rafters and the walls. Where two roof slopes intersect, as in an L-shaped house, valley rafters are added. These are doubled whenever the adjoining roof sections are approximately equal in size. Jack rafters run between the valley rafters and the ridge boards.

Prefabricated Trusses

Many new homes use prefabricated trusses instead of standard rafter construction. These use less material and are erected quickly in the field. Scientifically de-

Ridge board
Collar beam
End stud
Top plates
Ceiling joist
Rafter

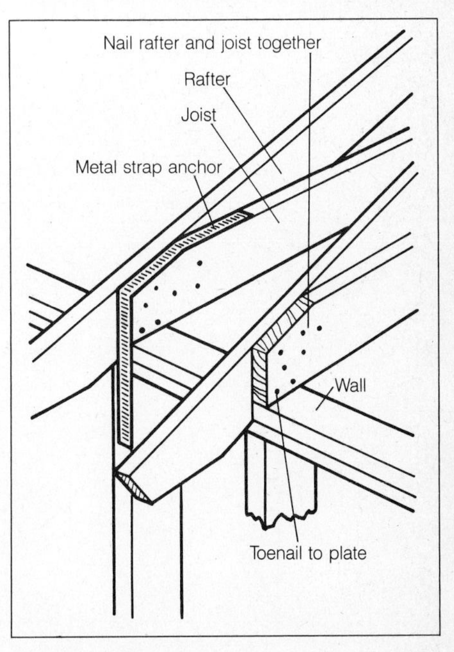

Rafter ends (and sometimes joists) are angled and nailed together and to outside walls. Strap anchors may attach joists to top plate.

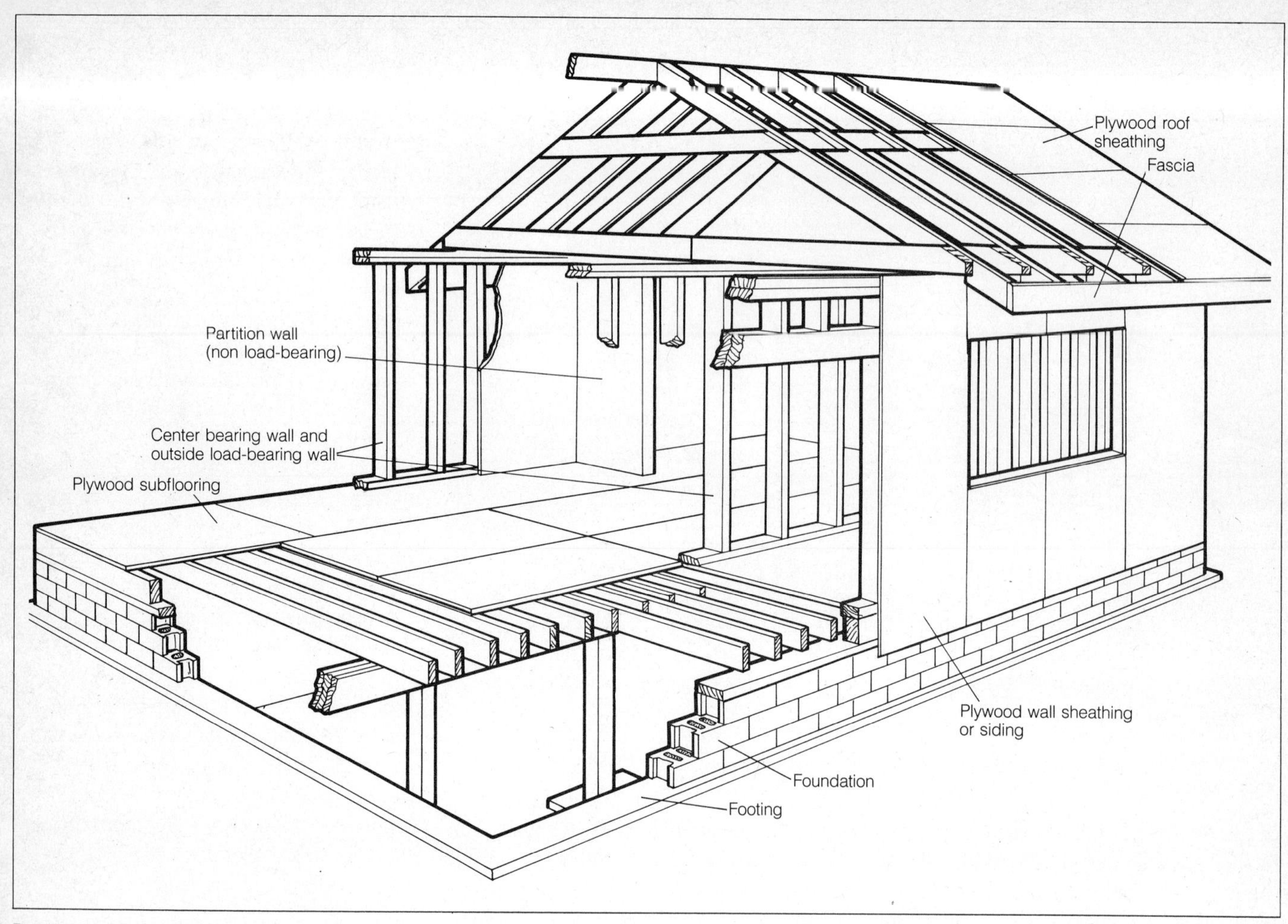

Decks are usually of plywood sheathing; however, spaced board sheathing may be used instead, particularly for wooden shingles.

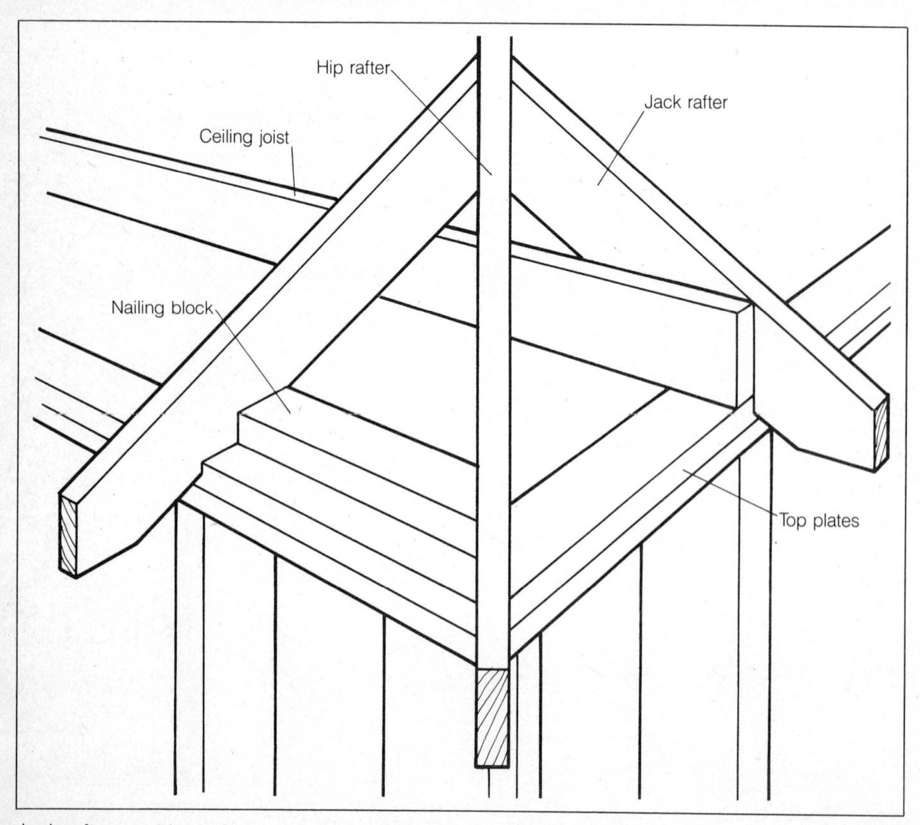

Jack rafters on hip roofs run between valley rafters and ridge boards, and between hip rafters and the walls, to help hold them together under stress.

signed to support loads over long spans without any intermediate support, they eliminate the need for interior bearing walls. Prefabricated trusses take advantage of the superior strength of triangular supports and plywood gussets. The most popular of several types of truss is the "W" truss. Other types are suitable for smaller spans and lower slopes, and the scissors version can be used when a cathedral ceiling is desired. Trusses attach to outer walls with strong metal straps.

Eaves, Cornices and Rakes

The word "eave" is used generally to refer to the outside, lowest edges of a roof. In the narrower sense, eave simply means the edges of the decking, and in some homes there is no more than that. Usually, however, you find a more complicated structure at the eaves called a cornice—an overhang. The cornice is the projection of the roof at the eave line; the projection forms a connection between the roof and sidewalls.

In gable roofs, there are cornices on

each side of the house, and in hip roofs the cornice is continuous around the perimeter. These cornices are usually formed by extending the rafters over the sidewalls, whereas cornices of flat and low-pitched roofs are extensions of the ceiling joists that also serve as rafters. In addition to the rafters or joists, which are the backbone of the structure, there usually is a fascia (face) board nailed to the ends of the rafters. The area underneath is called the "soffit". The soffit may be open, or closed in with plywood or other materials.

Although it may seem like a minor part of roof construction, cornices play an important role and are worth a close look. Many weathering problems, ice dams, and poor ventilation can be traced to poor cornice planning and construction. Unless you have a closed cornice, single-inlet vents or continuous-screened slots, vents should be used in the soffits. Here are some of the more common arrangements.

Close cornice. Rafters and ceiling joists are cut off at or close to the top

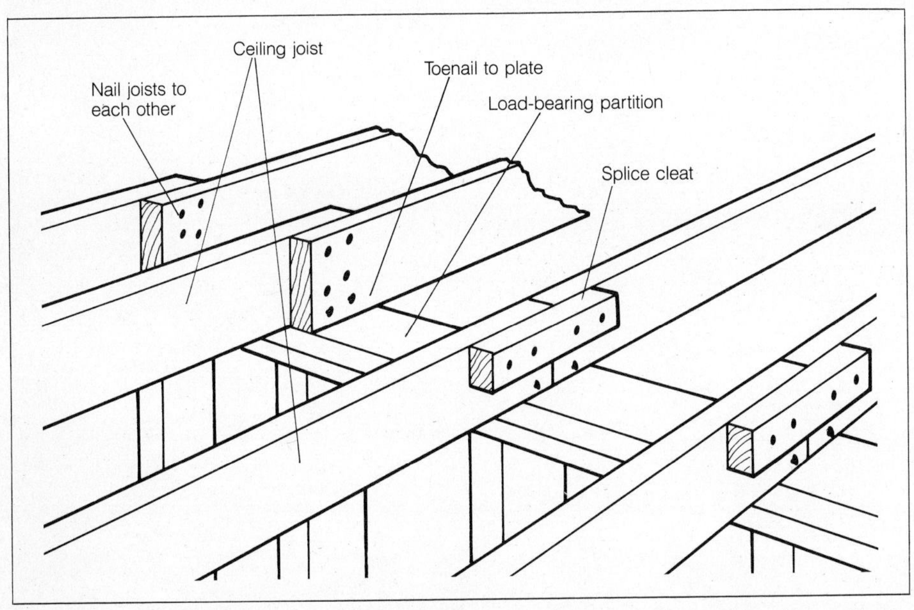

In typical home construction, ceiling joists are joined to an interior bearing wall by either nailing to ceiling joist, toenailing to plate, or splicing with a cleat.

Solid vinyl soffit system panels are one option when repairing old or damaged soffits; they offer fast installation and come in kits.

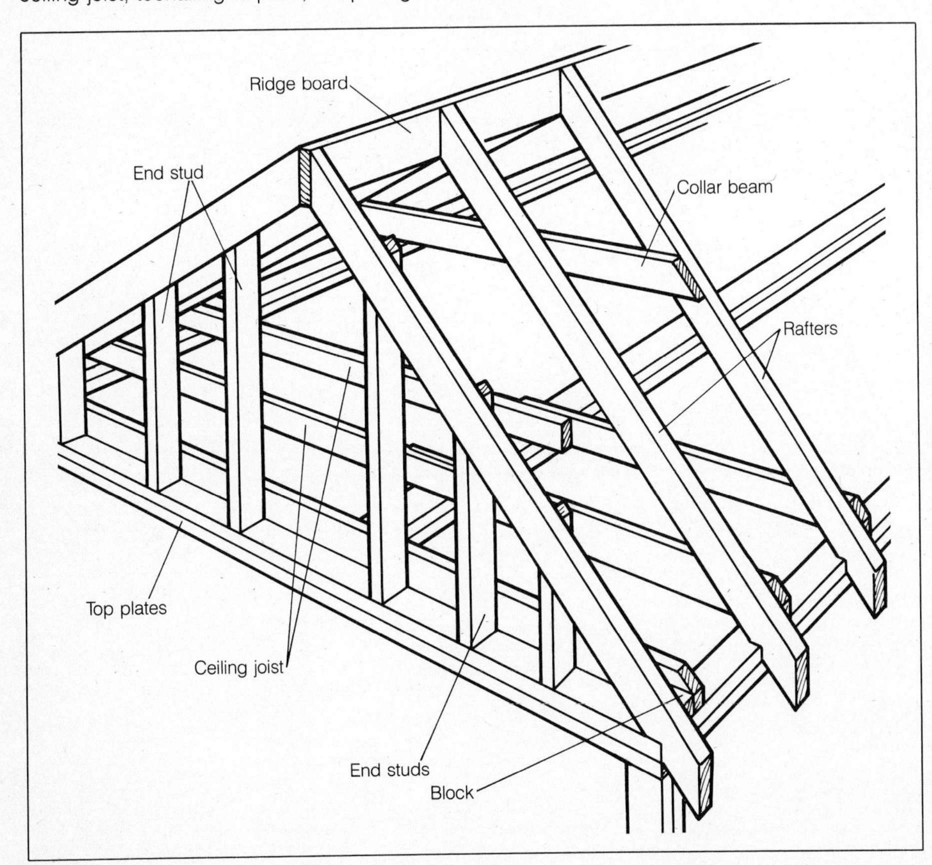

Shown is the completed roof framing. In this construction, blocks are used between rafters and joists. Note the collar beams placed between every third set of rafters.

Gutters and downspouts often are utilitarian and dull, but can be purchased in designs to match and enhance the home's exterior design.

Shown are 3 types of trusses. Version A (see above and below) is the most often found. The 4 "webs" form the W-shape. Version B is used for lower slopes (note the rises at the right). When a sloping ceiling is desired, as in a cathedral ceiling, the special scissors truss, C, is used.

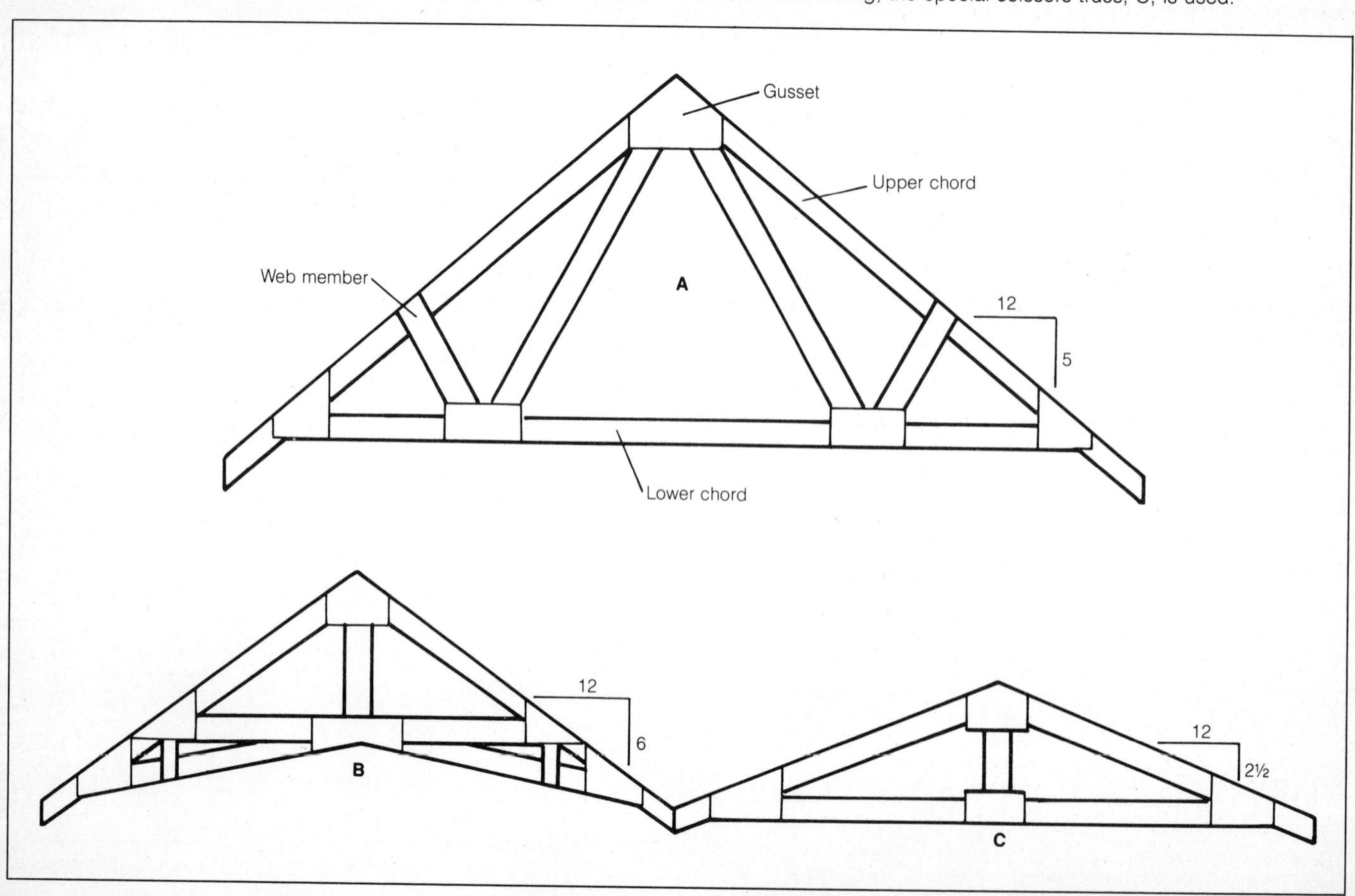

plates. The deck stops at the top piece of siding, and the shingles extend only slightly over the deck. Although the simplest (and cheapest) to build, this type of cornice leaves the roof with an "undone" look and provides little protection for the sidewalls. Even worse, there is no room for ventilators. An eave or cornice extension should be considered if you have this type, but the looks (if not the utility) can be improved by the addition of a formed wood gutter.

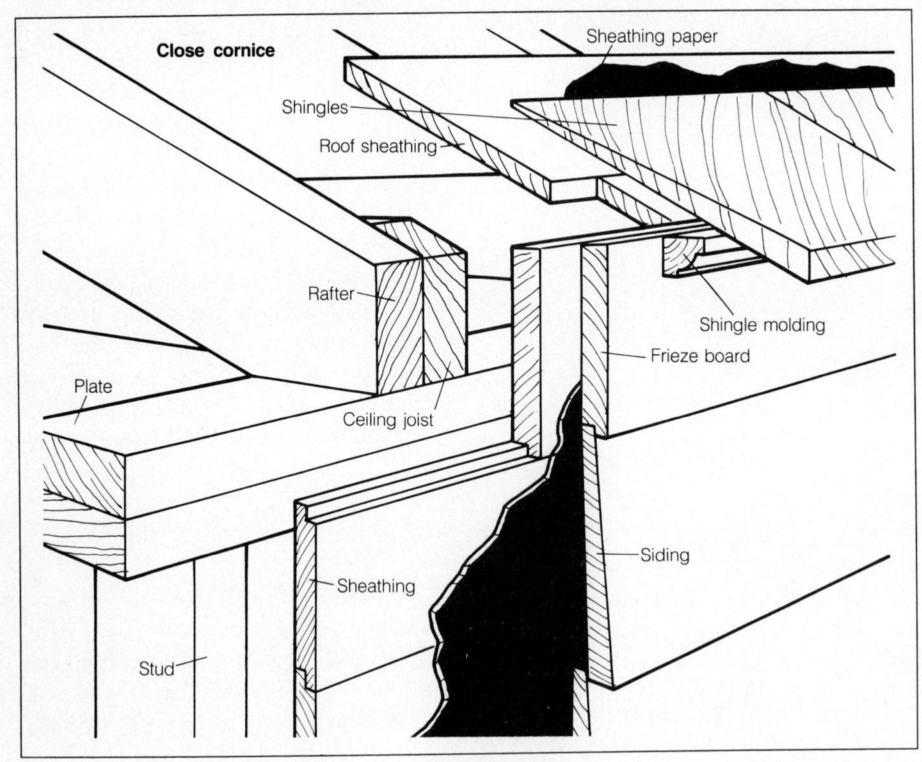

A close cornice is the least expensive and simplest type of cornice to build, but it does not offer protection to the side walls, or allow room for ventilation.

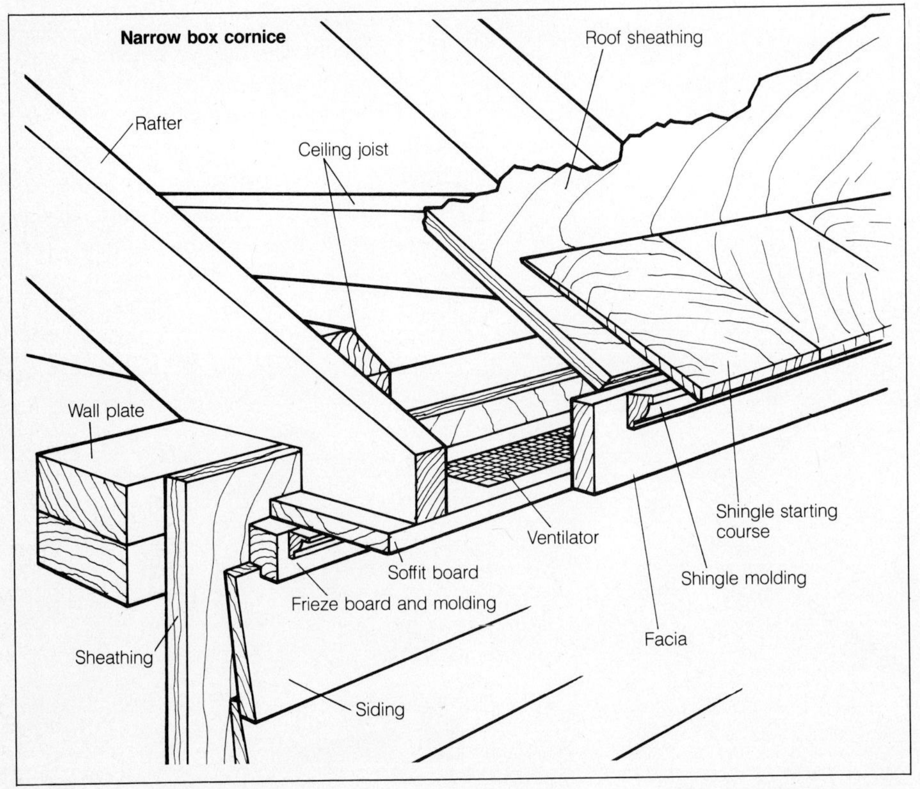

A narrow box cornice must be designed to allow sufficient room for placement of soffit vents for air circulation and ventilation.

Narrow box cornice. The projections of the rafters serve as a nailing surface for the soffit boards and the fascia trim. Depending on the roof slope and the size of the rafters, this type of cornice can extend 6 to 12 inches from the sidewalls. Although not as desirable in appearance or as effective for siding protection as wider cornices, it usually is adequate for ventilation. Replacement with wider cornices is generally uneconomical.

Wide cornice. A cornice over a foot, sometimes extending several feet away from the sidewalls, gives excellent protection to the sidewalls. Wide overhangs can even be used as a roof for a narrow patio or porch. Most hip roofs have wide cornices. Depending on the height and type of house, the wide box cornice may require "lookouts", horizontal framing members that are toe-nailed to the wall and face-nailed to the end of the rafter extension. In some homes, one-story homes in particular, the lookouts and horizontal soffits may bring the cornices too low. In that case, the lookouts are eliminated and the soffits are nailed to the bottom of the rafters, or the soffits are eliminated entirely for an open cornice.

Flat-roof cornice. Because the ceiling joists also serve as rafters, construction of cornices on flat and many low-pitched roofs is different from those of other roof eaves discussed above. Unless the overhang is very short, one of the two methods shown in the accompanying drawings is used.

One method, with an overhang of less than three feet, has joists extended in one direction, with lookout "rafters" (actually joists) used in the other direction, beginning with a doubled joist one or two joists in from the edge. The distance from the doubled joist is twice the overhang.

When the cornices overhang three feet or more, construction involves use of a doubled rafter to provide greater stability. Rafter ends are finished with a nailing header to permit fastening of the cornice soffit and fascia. Ventilation is necessary.

Sheathing and Decking

Once the rafters are all securely in place, the roof decking is applied as soon as possible, so that weather-free interior work can begin. Most modern decking consists of sheathing-grade fir plywood. The large 4x8 foot sheets go up quickly and are easy to nail down. All lumber and plywood used in roofing must be well seasoned to avoid warping.

Another type of decking used extensively in older homes is tongue-and-groove (sometimes ship-lap or square-edge) lumber. To prevent warping, deck sheathing boards should not be too wide. When wood shingles are used as the roofing, the decking is often of spaced one-inch lumber, with open areas left to permit drying of the wood shingles. Boards should not be wider than eight inches, preferably 1x6, to minimize shrinkage. When open sheathing is used with wood shingles, the boards should be 1x3 or 1x4, spaced the same distance on centers as the shingles are laid "to the weather" (dimension covered by the shingles in the course above). For example, with shingles laid five inches to the weather, nominal 1x4 boards will have their centers five inches apart. Since a 1x4 is actually 3/4 inch x 3 1/2 inches, there will be a 1 1/2 inch gap between boards for ventilation.

Although it is more important for wood shingles to receive direct ventilation than for other types of roofing materials, every roof should receive adequate ventilation. Attic areas can trap the moist air from cooking, showers and air-conditioning. When this damp air hits the cold roof sheathing, condensation results, eventually causing rotted wood and soggy insulation. To ensure good circulation, install louvers of sufficient size—placed high in the gables—or add vents to the eaves.

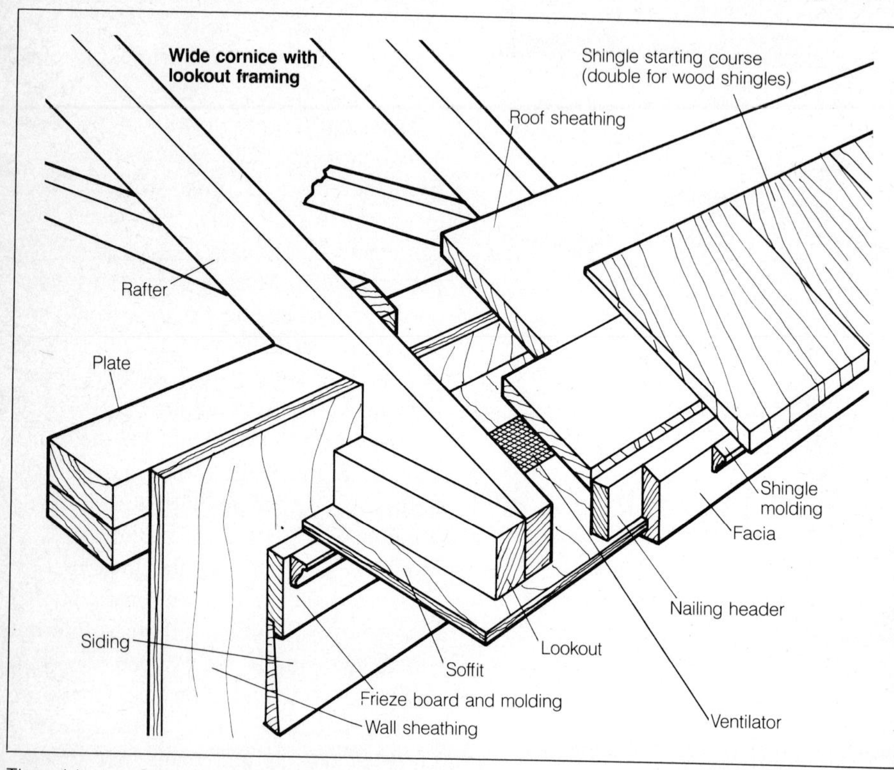

The wide cornice is boxed in with "lookout" framing. It usually is considered the most desirable type of eave construction, both visually and functionally.

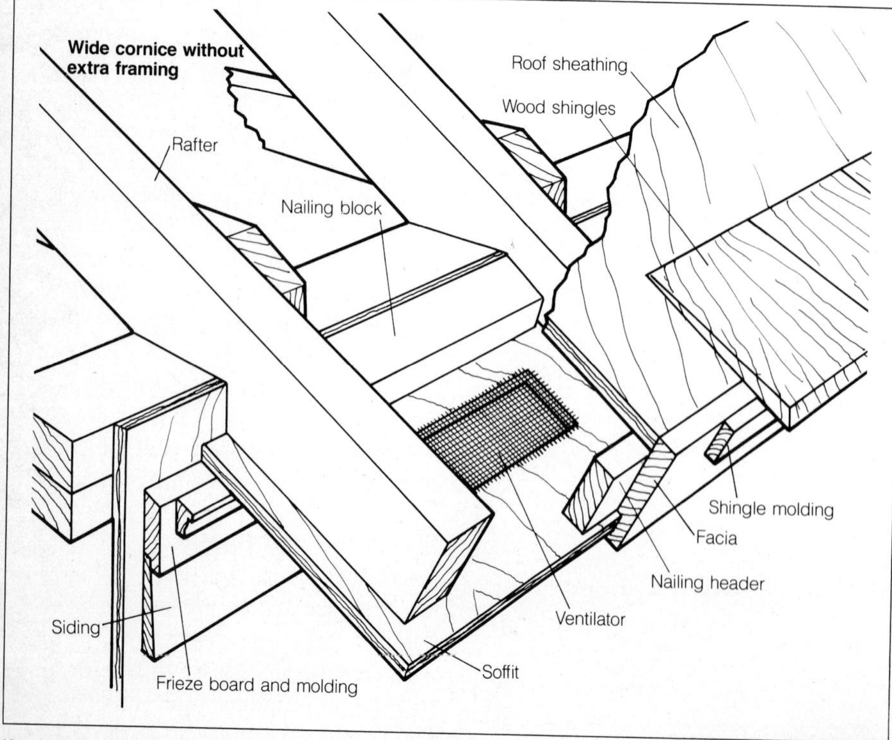

If a wide cornice with lookouts would extend too low, the extra framing is eliminated; the soffits attach to rafters. If the soffit is removed, it becomes an "open" cornice.

Shingles and Flashing

Once the framing is complete, the roof is covered with sheathing or decking, and then the shingles are applied. If you consider the framing as the skeleton, and the sheathing as the flesh, the finishing shingles correspond to the skin. Although the roof, like the body, can be dissected into various components, each must function as a complete entity. When one part breaks down, the entire "body" is affected. Therefore, a break in the "skin" can lead to decay in the sheathing or even the framing. If the framing is faulty or sagging, this can cause a rupture in the weathertight bond of the shingle skin.

An unusual architectural feature, such as the cupola shown here, sometimes can be covered with shingles to match the rest of the roof.

Poorly applied sheathing may allow seepage that affects the framing structure and/or the shingles. Incorrectly laid or flashed shingles can result in damage to decking, framing, and the house interior.

Exposure and "to the weather". Shingles are laid so that part of each shingle is exposed (this dimension is called the "exposure") and part of it is covered by the next course up. This distance that is covered by the next course is called the distance "laid to the weather". For example, if a shingle 8 inches deep is laid with 5 inches uncovered and 3 inches covered, it has an exposure of 5 inches and is laid 3 inches to the weather.

Flashing. Metal or asphalt flashing is installed wherever the shingles meet another plane, whether vertical such as a sidewall or a chimney, or at an angle for an adjoining roof deck. The flashing prevents water from seeping under the shingles. It must be kept in good condition if leaks are to be avoided.

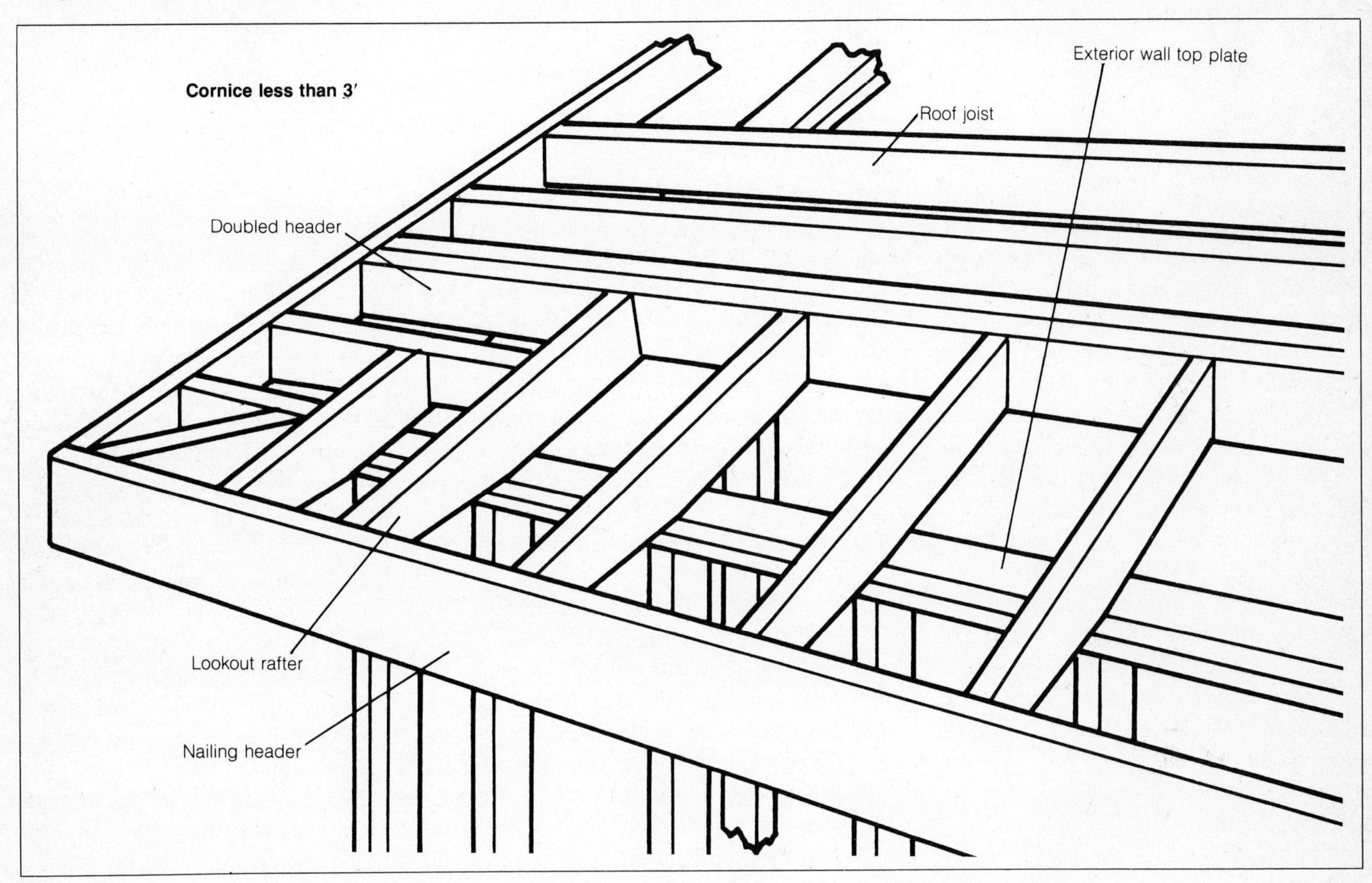

The corner construction of a flat roof with a cornice of less than 3 feet has "rafters" that really are extended joists.

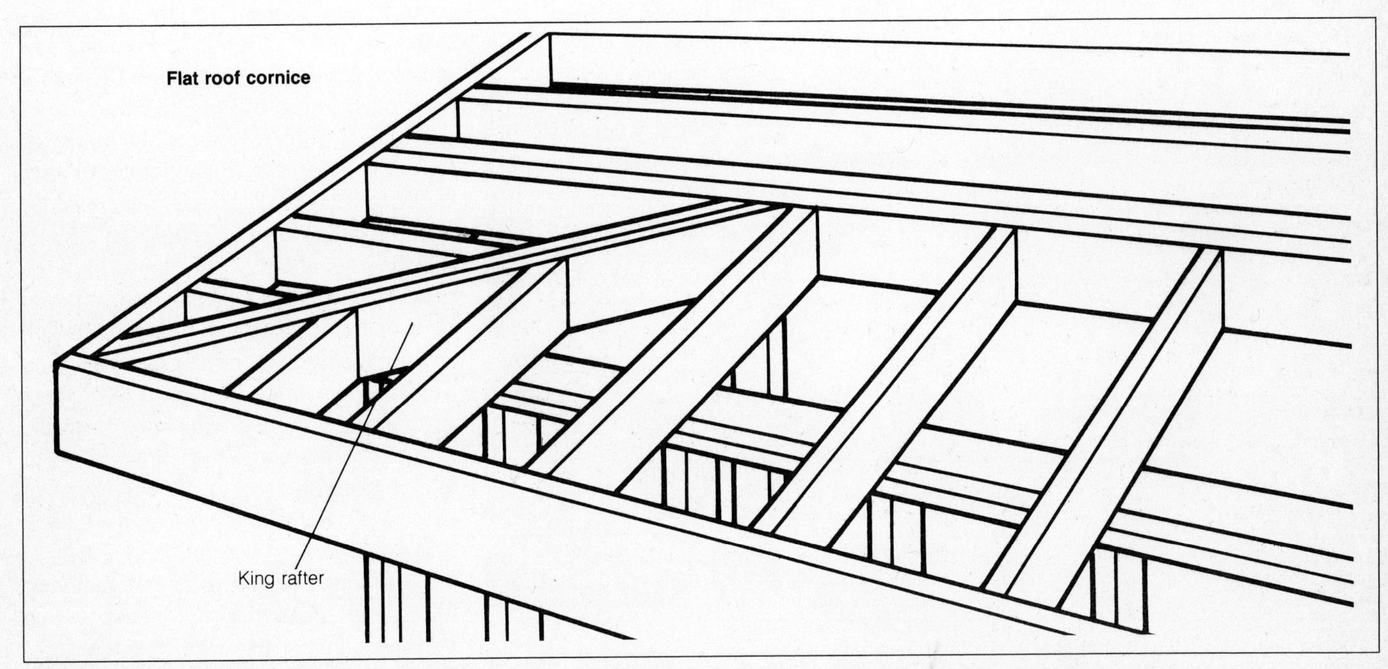

When a flat-roof cornice overhangs the endwalls by more than 3 feet, use a "king" or double-framed joist-rafter at the corners.

2

INSPECTING YOUR ROOF

Your entire home, including the roof, should be inspected once or twice a year for potential problems. Twice a year is best, once in the spring and once in the fall. For newer houses, however, a spring inspection may suffice. Even newer homes can have roofing problems because of shoddy workmanship or unusually severe weather conditions. In these cases, look for loose shingles or for leaks.

INSPECTING THE INTERIOR

In most cases, it will be necessary to go into the attic to inspect the interior of the roof. Exceptions would be rooms in one-story homes with cathedral ceilings, and houses with low, flat roofs such as those in the Southwest.

Access to the Attic

Although some attics are accessible by built-in or pull-down attic stairs, most homes built in the last quarter-century will have scuttle-holes.

Using a scuttle-hole. A scuttle-hole is a square hole in the ceiling of the top floor, usually in a closet. It is covered by a piece of plywood or scrap paneling, which ordinarily rests on pieces of molding around the hole. If you have never gone through a scuttle-hole before, there are a few instructions and guidelines you should follow when you try to enter one:

(1) find and set up a strong stepladder to get through the hole;
(2) if at all possible, get someone to hold and steady the ladder;
(3) push the plywood covering to the side, and stick your head through;
(4) do not do this in the summer if there is any chance you have a wasp infestation in the attic;
(5) wear a cloth mask if there is insulation in the attic.

Walking safely in the attic. Before you actually go up into the attic, take a look at the attic "floor." If yours is a typical attic, there will not be a floor at all. What may look like a floor under the joists or ceiling beams is the top of the ceiling material, probably gypsum wallboard. If you try to walk on this thin, light drywall, you will wind up with a leg or two dangling into the room below. If you have this type of ceiling, get some boards or ¾ inch plywood to walk on. How much of this you need depends on how big the attic is, and how much walking around you will have to do. Make sure you have enough; err on the side of too much rather than too little. If you are using plywood, cut 4x8 sheets into two 2x8 halves so that you can get the wood through the hole.

Before you attempt to go up through the scuttle-hole, push the boards up into the attic on top of the joists, far enough away from the hole so that you will have room to pull yourself up. Bring a screwdriver or other probing instrument, and either a flashlight or a trouble light on a long extension cord. Place these on one of the boards before you get up there, or have a helper ready to hand them to you.

There may, of course, be a floor already laid over the ceiling beams, in which case none of the above will be necessary. Some attics may have a partial floor or walkway already installed, which should be sufficient for inspection purposes. However, it won't hurt to have some extra boards in case you have to probe into one of the corners. Be careful about walking on this type of surface; some of the boards may be weak, or not nailed down.

Now that everything is in readiness, climb high enough on the ladder so that you are about waist high into the attic. Rest your forearms on the floor boards or

Some scuttle holes have ladders that are incorporated into the ceiling structure, while others are just a piece of wood that must be pushed aside.

the joists next to the hole, and pull yourself up into the attic. Once there, keep your weight on the joists if there is no floor, and push some boards into position so that you can walk or kneel on them. If the attic ceiling is low, you will probably have to crawl around on your knees, except at the center of the attic. If you can stand in the center, try to conduct most of the inspection from there. Otherwise, be prepared for sore legs and knees. Do not get your head too close to the inside of the roof sheathing. Very often, the roofing nails protrude into the attic, and they can give you a nasty gash in the scalp. Consider wearing a hard hat if this seems to be a problem.

As much as possible, try to line up your walking boards so that the ends rest just a little bit over the edge of a joist. If they protrude too far over beyond the joists, you may step on the unsupported ends and fall through. If the board ends are not overlapped enough, and do not rest solidly on the joists, they may slip off—again, a dangerous situation. You don't want part or all of you to end up in the room below.

When using boards or plywood as a walking surface in the attic, make sure the ends of the lumber extend far enough past the joist to prevent tipping or slipping through the joists.

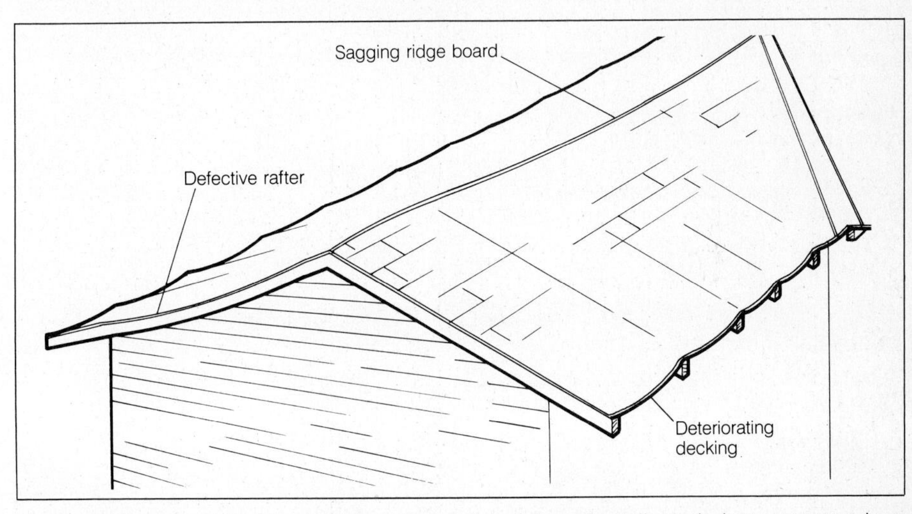

Shown are some serious structural flaws that demand "near future" repair, in many cases by a professional. These defects usually can be seen from the ground.

Trouble Signals

Once safely up into the attic, look carefully at the undersides of the sheathing. Sheathing is usually plywood, but it also can be made of spaced tongue-and-groove boards. The underside should be dry and uniformly aged, even after a rain, and there should be no dampness or discoloration of the wood. Either condition suggests at least a minor leak, which could lead to major troubles in the future. Pay special attention to the places where the roof is penetrated by a chimney, vent pipes or other structure. Check all around these openings for evidence of leaks, which is a strong indication that the flashing around them needs repair.

Wood rot. Examine the sheathing and the rafters for evidence of wood rot. Severe cases will be evident on sight, since the wood will be damp, dark and/or show signs of crumbling. Probe also with the screwdriver where there are discolored spots. If the tool goes in easily, the wood is soft and probably is moisture-damaged. Make a note to replace any deck material (sheathing) or rafters that are in poor condition. Soggy wood often occurs at the eaves, so check these areas carefully. You may have to stretch out on your stomach to get to these points, since the headroom will be very low.

Finding the leak source. Sporadic moisture inside the attic usually means there is a defect in the outside roofing material. Sometimes defective roofing will be directly above the discolored or damp wood, but not always. The water can get under the shingles in one place and travel some distance before it gets trapped and seeps inside. If, for instance, the tell-tale discolored area is at a joint in the wood deck, this may mean that the water seeped under the roofing until it found a point of easy access between the boards. For the same reason, it is even more dubious to assume that a leak in a room ceiling below is directly above the damaged area. In this instance, the water travels an even longer path from the faulty roofing through the decking and down onto the ceiling. The wet ceiling may be quite a distance from the source of the leak.

The discolored or damp sheathing is a good place to begin your search. To aid the detective work when you get out later, push a wire or long thin nail up from the attic through the deck and the shingling. This will mark the spot and help you locate the suspected area from the outside without guess work. Then you can try to trace the leak back from the roof surface.

Humidity problems. If air in the attic

seems especially humid, the problem may lie with the ventilating system not with the roof itself. Overall moisture problems are caused by an insufficient number of vents or by the lack of a vapor barrier. The vapor barrier is on the bottom of the insulation, facing toward the room (or it should be). Lift up the insulation to see if there is a foil or paper covering (vapor barrier) on the underside. All houses should have some insulation, so resolve to install it if you don't have any. If the insulation is the ''pour'' type, sheets of polyethylene should be used as a vapor barrier.

Danger: bees, wasps, bats and squirrels. Bees and wasps like to build nests in attics, and squirrels sometimes find a way inside for gamboling on the joists. Bats may also have found a home there, usually in older homes. If you have such unwanted guests, you probably know it in advance, but it pays to be cautious.

Ordinarily, none of these species will bother you if you don't approach them. You may, however, stumble on them unexpectedly, surprising them as much as they surprise you. Bee and wasp stings are painful at best, and potentially lethal for those with allergies to their venom. Bats and squirrels may harbor rabies.

To avoid encounters of this unpleasant kind, look carefully before you stray into unfamiliar territory. Insect nests are easily spotted. Squirrel and bats will usually make their presence known, but survey the area carefully with a light before venturing into it. Bats will rarely make a move in daylight, so don't make an evening call if they are suspected. Since an animal normally will run the other way if it sees you coming, avoid trapping it in a corner.

If you have an unfinished attic with loose fill insulation, this is probably what you will first see when you put your head through the scuttle hole.

Trouble signs sometimes can be seen without ever going up on the roof—such as this loose and warped asphalt shingle, which would be clearly visible from the ground.

When There Is No Attic

If you live in a house with a flat roof, a cathedral ceiling, or with an attic that is inaccessible or too low to crawl around in, interior inspection is difficult—if not impossible. The only evidence available from inside is an obvious leak, at which time it is too late for preventive steps. You are now into the area of repair, unless you happened to catch the leak right away, before any permanent damage occurred to the soft ceiling material.

EXTERIOR INSPECTION

When interior inspection is impractical, external inspection and maintenance are doubly important. Such roofs must be inspected at least twice a year from the outside. Since flat roofs are made of built-up roll roofing, much as a commercial roof is, professional roofers should ordinarily be consulted for both inspection and repair. This is especially important when there are layers of stone, gravel or other materials on top of the roll roofing. This layer should not be disturbed by inexperienced feet. If your roof is made of built-up roll roofing without any mineral or stone overlayment, there are some routine repairs that you can do yourself, as covered in Chapter 6.

A proper roof inspection is more than a casual task. Unlike a car, you cannot just ''raise the hood'' and look inside. For some roofs, such as those with many gables or dormers or which are high off the ground, the following safety preparations are mandatory.

Tools and Clothing

Start with footwear. No matter what type of roof you have, you should wear something with rubber or crepe soles. This type of footwear not only provides an extra measure of safety against slipping, but it also helps prevent unnecessary damage to the roofing. Workshoes with strong ankle support and nonleather soles are best. If you don't have these, wear regular shoes with nonleather soles. If nothing else is available, a good pair of high sneakers will do. Low-cut tennis shoes are not recommended because of their poor ankle support.

The best way to inspect the exterior of your roof is to go up there and take a close look. You will need a ladder for that; the type depends on how high the roof is, how steep the pitch, and how accessible it is. A simple, low, one-story house can sometimes be reached by a high stepladder. More often, a low straight ladder will be required. Two-story houses cannot be inspected without the use of an extension ladder. If you have a high townhouse, or any other home higher than two stories, scaffolding may be necessary. For homes that high, it may be better to hire professional roofing experts to perform the inspection, although it may be possible to attempt at least a cursory job with binoculars or a telescope. (A telescopic camera lens, if you have one, may do the job.)

A low-sloping, one-story roof can be inspected without any additional harnesses or ladders other than the one needed for access. For steeper pitches, you may also need "chicken ladder" and/or a rope safety line or harness. See Chapter 5 for the details of this equipment and other safety equipment. You will have to make your own judgment on this, evaluating the type of roof and your own confidence about climbing around up there.

The most important tool in roof inspection is one that nature has already given you—your eye. If you need eye-glasses, be sure to wear them or bring them. A roof inspection will be both long and close range, so don't forget glasses, even if you normally need them only for close work.

You will also need a tool for probing wood to find moisture damage. If you have a carrying sheath for a scratch awl or an icepick, these are your best choices. However, a screwdriver is almost as good and it won't dig a hole in your leg as you're climbing. This is really the only tool you must bring with you, but you may also want to bring some repair tools and equipment if it's a difficult job getting up and down. That way, you can make any simple repairs while you are performing the inspection, avoiding another difficult climb. Another possibility is to take this opportunity to clean out and repair the gutters while you are up there.

Ladders. No matter what type of home you live in, you will need some type of ladder to reach the roof. Before attempting to get up on top of the house, read Chapter 5 regarding the selection and use of ladders. The misuse of ladders is one of the most common causes of injuries in the home. Just as an aside to this, and even though there is no objective proof, the author personally feels safer on a wooden ladder than on an aluminum one. The wooden ladder seems more solid, it does not sway as much, and it cannot conduct electricity if you happen to stray near live wires. Experienced roof climbers will undoubtedly scoff at this, pointing out that wooden ladders are much heavier and harder to carry and set up. (You can get hurt if a heavy ladder falls on you, too.) Everyone knows, they will point out, that you would be stupid to put a ladder where it can contact the wiring. Still, if climbing a ladder makes you nervous, or if you are at all afraid of heights, I think you will prefer the wooden variety.

Always check out a wooden ladder first to make sure that all the joints are intact and that the rungs are solid (see Chapter 5). Wooden ladders are more susceptible to weather damage and to aging, and do not last as long as aluminum ladders. The older the ladder is, the more necessary the inspection.

Going onto The Roof

Your view from the ladder may indicate that the roofing seems to be basically in good shape. However, before descending the ladder, at least inspect the flashings around the chimney and other protrusions. Even if conducting an annual inspection, rather than looking for specific leaks, you also should check underneath the shingles and along the flashings.

If the roof is very steep, you should have taken one of the precautions suggested in Chapter 5. If at all possible, put your "chicken" ladder or rope sling in place from the ladder. If this is not feasible, nail a piece of 2x4 to the edge of the roof where you can step onto it easily from the ladder. Use this as a firm footing while you inspect the rest of the roof.

Set one foot firmly onto the roof and hold onto the ladder, shifting your weight onto the first foot as you move the other onto the roof, letting go of the ladder. Whatever you do, don't step on the gutters. They are intended to carry rainwater, not your weight.

What to Look For

Once safely on the roof, you can move cautiously about, checking the shingles for warping, cupping, looseness or rot. Examine the roof for any shingles that may have been lifted by the wind. You may find that some shingles are missing. Bumps indicate that the roofing nails below have become loose. Look also for any nails that have popped right through the shingles. If this situation is common, the entire surface probably needs replacement.

Raise extension ladders one rung at a time, by hand. Some ladders have rope/pulley setups. Check that the hooks are locked.

Open and lock the ladder. Lower the paint tray holder on the front of the ladder to further lock the ladder in position.

The condition of a wood roof may be difficult to judge from a distance, and probably will require a trip to the roof.

If the damage to the wood shingles is localized rather than widespread, replace the damaged shingles as shown in Chapter 6.

Wood shingles and shakes can be the most beautiful of roofings, but not if they are old, rotten, warped, and split.

Slate is easily recognized, although it usually changes color over the years and often chips at the edges.

Metal tile is not often seen on residential roofs because it can become rusted and corroded, may attract lightning.

Repair of an asphalt shingle roof on which the decking and framing are in good condition can be attempted by the homeowner.

Leaks. If you have found evidence of leaks on interior inspection, and have inserted a wire or nail through the roof, try to determine the source of the leak. If the shingles immediately around the wire are missing or loose, the problem is obvious. Otherwise, raise up the shingles in the area, pulling out the nails if necessary, and try to trace the leak to its source by following the discolored trail in the roof decking.

Wood shingles. Evidence of wood shingle deterioration is indicated by discoloration along the edges. Individual shingles that are considerably darker than the other should be suspected and tested for rot with a screwdriver. Replace cupped wood shingles that have been alternately subject to excessive wetness and drying out. Occasional cracked shingles can be repaired, and new shingles can be individually substituted for old warped or broken shingles.

Built-up roofs. Built-up roofing on flat or low-sloping roofs should be examined for bare spots on the surfacing and for separations and breaks in the felt. Bubbles, blisters or soft spots indicate that replacement or major repairs are in order, although you make temporary repairs to put off the evil day. Alligatoring (cracks and blisters) on a smooth-surfaced roof may or may not be a sign of major trouble; it could be just a dried-out surface, and a coat of roofing cement may be all that is needed. Check for looseness of the mineral granules at the surface. If they come out easily, the asphalt-impregnated felt that holds the granules has dried out. You may wish to consult a roofing contractor for an additional evaluation.

Checking the flashing. "Flashing" refers to the sheet metal or roll roofing installed at crucial intersections of the roof around chimneys or vent pipes, and in valleys where two different planes meet. There should also be metal or vinyl

Junctions of planes, or between the roof deck and vertical obstacles, are prime suspects for damaged shingles and flashing.

drip strips at the eaves and rakes, although this is not technically considered part of the flashing.

Examine metal flashings for loose pieces or for corrosion. These conditions are almost certain to cause leaks. Take a close look at all the places where flashing is found. Make sure that flashings are in place at all junctions of the roof, and determine if any need repair or replacement. Where possible, lift up some of the shingles where flashings should be located underneath and check their condition. Flashing repair in most locations is simple, but it is extremely important.

Inspecting eaves and rakes. Shown are the locations of rakes and eaves on a typical house. Basically, these are terms for the edges of a roof. The eaves are along the low horizontal edges. The rakes are the ends of the roof on top of gable ends. On a simple gable roof, there are two eaves and two rakes. The main part of the roof in the accompanying art is a hip roof, with eaves on all sides except for the gable addition in front.

Both eaves and rakes should extend at least a few inches beyond the siding. Consider adding an eave extension if the overhang is small. The extension gives protection from water drainage and often contributes to the appearance of the house exterior.

Due to ice dams, faulty gutters and the fact that all rain and snow eventually runs over the eaves, the fascia or cornice board nailed to the end of the rafters—and to which the gutters are attached—often suffers damage. This, in turn, causes the gutters to loosen because the wood will no longer hold the nails or straps. The same condition may be found in the wood soffits that form the underside of the overhang.

The edges of the deck, or eaves and rakes proper, are best inspected while still on the roof. Lift up a few shingles (carefully) to check the condition of the drip edges nailed into the ends of the deck. Where possible, probe into the wood under the metal or vinyl drip edges. If you find evidence of sogginess there, but the rest of the deck seems in good shape (not an uncommon condition), you may have to replace the decking at the edges only.

The roof overhang, if any, will consist of an open or boxed cornice. If you have an open cornice, with no soffit boards

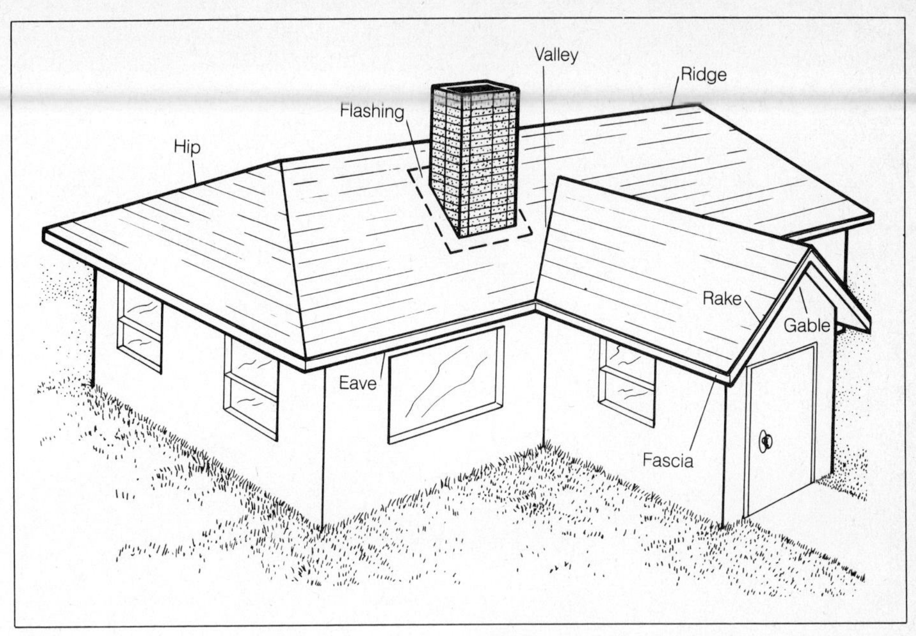

Shown are some of the most common locations of roof deterioration and leakage. Some of these points such as the eaves and fascia, can be inspected from a ladder.

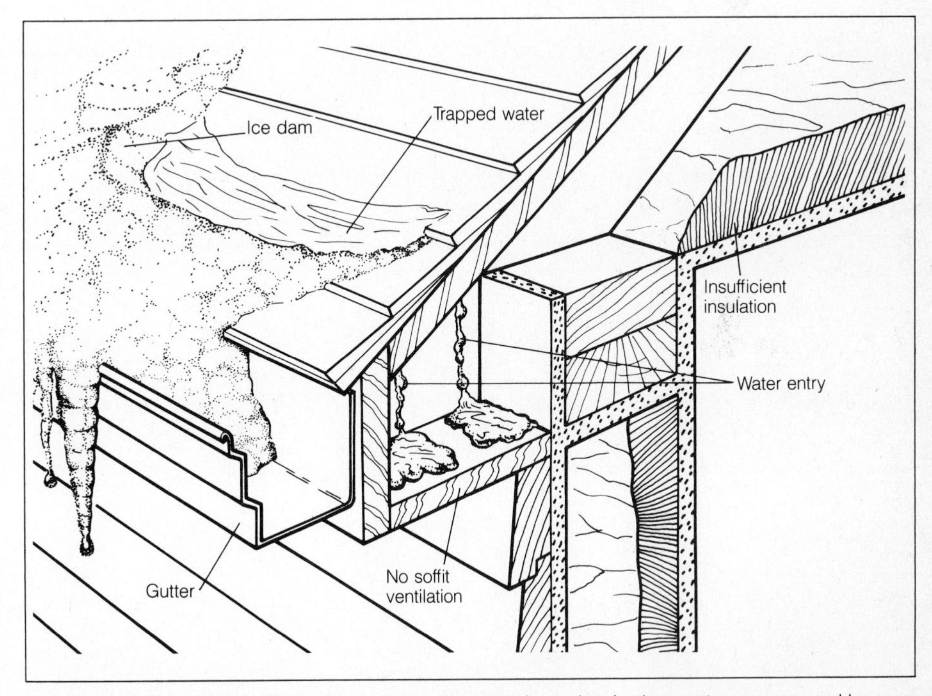

Ice dams, which can result in serious interior damage due to backed up water, are caused by poor ventilation and insulation.

Rafters, eaves, and other wooden members will soften and deteriorate if subject to moisture over a long period of time. Probe for wood rot.

under the rafters, you can inspect the deck from underneath. If it's safe to do so, lay down on the roof and probe into the sheathing and rafters from below. Don't ever try this on a steep roof, however, or you may slide off. Work from a ladder to carry out the same test.

You won't be able to get at the deck or rafters if you have a boxed cornice, but inspect the cornice area in any case for evidence of wood rot. Move the ladder around to various places along the eaves; check all the soffits, watching particularly for clogged vents. Make sure that they are all screened and they are adequate for the amount of attic space.

Eaves similar to the one shown, with peeling paint and old wood, often need more than just a new coat of paint. Replacement usually is suggested.

In areas of severe winds or heavy and strong rainstorms, shingles may be blown to other parts of the roof or lifted completely off the roof.

Gutters and leaders. If there aren't any gutters along the eaves, there should be. Make a note to install them if they either are missing or inadequate. As long as you are on the roof anyway, you may as well clean out any leaves or other debris that prevents the rainwater from running freely. If leaves are a problem, plan to install screens over the gutters.

Gutters should empty into "leaders" or "downspouts." There should be ball-type strainers at the top end of the leaders. If there aren't, you can buy them at most hardware stores. They are well worth the small cost.

You should, of course, note the condition of both gutters and leaders. Look for holes and corrosion; if you find any, repair or replace the damaged units. Make sure that joints are solid and leak-free, and that gutters are properly pitched toward the downspouts. See Chapter 10 for more on gutters and leaders, and their repair or replacement.

When Not to Go Up On the Roof

No matter how brave you are, don't attempt to go up on the roof at all on windy days, or if there has been a recent rain. Be sure any breeze is calm and that the roof surface is absolutely dry. Never go near any power lines, and disconnect the television set if you have a roof antenna. Even then, stay away from the antenna because some antennas retain an electrical charge even when the power is off. If this is a problem, disconnect the antenna at the back of the TV as well as pulling the plug. It is also recommended that you read Chapter 5 and absorb all the safety tips there.

Fear of heights. Anyone with even a slight fear of height (acrophobia) should not attempt to go up on a roof over one-story high. Severe acrophobiacs shouldn't even attempt that. Even if you conquer your fear momentarily, you will be too nervous to do a proper inspection job. Have someone else in the family carry out the inspection for you, or hire a professional.

3

CONTRACTORS VS. DOING IT YOURSELF

One way or another, you have now completed your roof inspection. If you're fortunate, that's all you have to do until next year. Almost as fortunate are those who discover only an occasional lifted shingle, or perhaps a minor, easily repairable leak. These simple and inexpensive tasks are presented in Chapter 6. It is gratifying to find that the roof discovery in general seems in good shape: the mineral granules are tight which means that the asphalt-saturated felt is still supple, and has not yet become dried out and brittle; there is no major lifting, curling or cupping of the shingles; the deck and framing are all dry and solid.

Some of us will not be so lucky. Once it is obvious that major repair or replacement is necessary, a number of questions arise. How much will it cost, and how will I get the money? Can I do the work myself, or should I hire a contractor? Or can I work out a combination of both? If I hire a contractor, how do I find one who is reliable? What sort of roof should I put up? Must the old shingles be removed? The questions seem endless.

Eventually, even the best of roofs deteriorate, no matter how well maintained. Sooner or later, a roof reaches the "point of no return." Wear and tear will take its toll, and you will either need an entire new roof, or substantial renovation. In some cases, not only the exterior roofing materials, but also the roof deck and perhaps even the framing will need repair. The trick is to know when a minor repair will suffice, or when reroofing is required.

WHAT NEEDS TO BE DONE?

What is the extent and magnitude of the job? If there is general deterioration, the entire roof should be reshingled. There

HOW BIG A JOB?
To "How big is the job?" one may well respond "How big is big?" A lot depends on skills, availability of time and materials, the accessibility of your roof and similar factors. Let us take as our criteria for "minor repair" that it can be performed on a not-too-hectic day in about four to six hours of steady work. Here is a list of most roofing jobs, classified as either "major" or "minor."

Major
- Complete reroofing
- Partial reroofing
- Deck replacement
- Gutter replacement
- New framing
- New chimney flashing
- Building a dormer
- Eave extensions
- Creating a skylight

Minor
- Fixing loose or damaged shingles
- Replacing missing shingles
- Fixing most leaks
- Deck repair
- Eave repair
- Gutter repair
- Framing repair
- Flashing repair
- Installing vents

Repairs probably should not be attempted by a homeowner when the pitch is as steep as on this house. In such cases, call in a professional.

The vinyl soffit and downspout system shown comes in a kit and is priced to encourage replacement without aid of a contractor.

Wood shingles and shakes, preferred because of their beauty, require slightly more complicated procedures than called for with asphalt shingles.

A mission style roof such as this one is low enough and accessible enough for a safety-conscious homeowner to undertake his own repairs.

are several corollaries to that decision, however. First, you must determine what materials to use and what type of roofing is already there. If you don't know what type of roofing you want, read Chapter 4. The best choice is usually either asphalt or fiberglass, which is the same except for the backing material, or wood shingles. Unless the housing is definitely of Mediterranean design, clay tiles are rarely used, and even more rarely installed by the homeowner. Asbestos shingles aren't made any more, and slate is extremely rare and expensive. If you're planning on doing it yourself, asphalt or fiberglass shingles are the overwhelming favorite, with wood shingles or shakes an attractive—but expensive and more difficult—second. Aluminum "shingle-shakes" are a relatively new option and may be difficult to find. If considerable structural work needs doing, and you know little or nothing about carpentry, let a contractor do at least that part of the work. It is not a bad idea, in any case, to have a few estimates done by professional roofers, so that you can get a good idea as to what the job costs. This will help you make a more informed judgment as to whether you want to do all, part, or none of the work yourself.

Partial Deterioration

Your inspection may have shown that only part of the roof has undergone serious deterioration. Sometimes, the eaves and rakes are in much worse shape than the rest of the roof. Often, that part of the roof facing the wind will need replacement, with the rest of the shingles still reasonably intact.

Deck work. It may be that you have determined that most or part of the deck has succumbed to wood rot. The best course is to tear off the old deck, in whole or in part, and replace it with new sheathing. Deck replacement is not only difficult, but can be very dangerous, because there will be only rafters to walk on after the old deck is removed. Depending on the dangers involved and your carpentry skills, it often is wise to hire a contractor for this part of the work. It is possible, however, that your roof framing is strong enough to support a new deck laid over the old one. Some roofing materials can be put on top of others; some cannot. The condition of the present roofing, as well as the number of layers of roofing already applied, must also be taken into consideration.

Structural repairs. Structural deterioration, of the rafters or other framing members, can usually be fixed from inside by doubling the existing rafters. However, if the ridge board or rafters sag, more drastic structural repairs may be in order. This problem, although rare, should be fixed by an experienced contractor. In most cases, the defects can be corrected without replacing the entire framing, but that possibility does exist.

DOING IT YOURSELF

Do not be frightened off by the discussion above of more drastic repairs. Unless you have just bought a ''Handyman's Special,'' which hasn't seen a hammer or paint brush in decades, very few roofs need repairs this difficult. Most roof repairs are simple and involve only the application of some roofing cement, replacement of a few shingles, or perhaps application of some sheet metal or caulking. Chapter 6 discusses these in detail and, if that is all your roof needs, you can proceed to Chapters 5 and 6.

Can You Handle It?

No matter how large or small the job is, there are certain value judgments you

Construction of gable dormers is complicated. Do not expect to do this work unless you have had previous carpentry experience.

have to make about your own abilities and skills. Perhaps even more important are such intangibles as how you feel about working on a roof. If you feel faint climbing a stepladder or looking out of a second-story window, do not even consider inspecting the roof, much less repairing it. When the roof is a high one, and the job involves walking on bare rafters, you must seriously consider the safety factor, no matter how brave you are. Your widow (or widower) may not appreciate a new roof if she (or he) has to live under it alone.

Skills required. Take into account the time and patience required. Shingling is a time-consuming, exacting, and boring job. If you are the type that likes to attack a job with vigor and get it done quickly, roofing may not be a good job for you. There is good news, however. Roofing is not a job that requires a great deal of technical finesse. If you can use a ruler and a hammer, you should be able to lay asphalt or fiberglass shingles with comparative ease. Wood shingles take a little more skill, but the techniques are easily learned. The basic requirements are those three intangibles—confidence in working off the ground, patience and attention to detail.

Help from the manufacturers. The majority of shingle manufacturers direct their marketing toward roofing contractors more than to do-it-yourselfers. One exception is the Celotex division of the Jim Walter Corporation, which has recently come out with a do-it-yourself kit for the homeowners. You can buy the kit for $10, try out their system and, if you decide to do the whole job yourself, the price of the kit is deductible from the overall cost if you purchase their shingles. Write to Celotex at P.O. Box 22601, Tampa, FL 33622, if there is no Celotex dealer in your area. Several other companies, listed in the appendix, provide information sheets.

SELECTING A CONTRACTOR

Let us assume now that you have decided that you need a contractor to do at least part of the work, perhaps all of it. The next problem is how to find a good one. Unfortunately, not all contractors are good, and not all of them are honest.

Perhaps the worst way to select a contractor is to hire one that comes to your door, claiming that he wants to make your house a ''showcase'' for his work in the neighborhood. He will give you a

For steep roofs, various arrangements can be used to provide a surface from which to work on the roof, which at the same time minimize the risk of slipping or falling.

ROOFING OVER OLD MATERIALS

New Material	Asphalt or Fiberglass	Wood Shakes	Wood Shingles	Clay Tile	Slate	Asbestos
Asphalt	yes	no	yes*	no	no	no
Fiberglass	yes	no	yes*	no	no	no
Wood Shakes	yes	yes*	yes	no	no	no
Wood Shingles	yes	no	yes	no	no	no
Clay Tile	yes	no	yes*	no	no	no
Slate	yes	no	yes*	no	no	no
Asbestos	yes	no	yes*	no	no	no

**Special procedures necessary.*

Cutout notches that separate each shingle into tabs (usually 3 per shingle) are centered on courses above and below.

substantial discount if you let him do your house and put his sign up. Time is always of the essence to this type of contractor (a salesman, really) and you must sign on the dotted line quickly for his boss to approve the contract. Often, he will "highball" you, by quoting a price obviously too high. When you resist, stating that you'd like to have it done, but that the price is too high, he will contact "the boss," and reluctantly quote you a price considerably lower. Now that your only defense has been shattered, you sign up without making any other comparisons.

This type of wheeling and dealing is more common with siding contractors, but often the same firm also does roofing. He may snag you for both, if you don't watch out. Some, as a matter of fact, do a pretty good job, and sometimes the price is not bad. But never sign up for any major improvement such as this without checking references and comparing several estimates.

Bids and Contracts

No method of choosing a contractor is foolproof. Get references from previous customers, preferably those who have had their roofs for several years. (A roof put up last month will be good, no matter how bad the application.) Find out how many years the contractor has been in business. Check out the Better Business Bureau and your local consumer agency for complaints.

Always get at least three estimates, and have the contractor specify the type, weight, manufacturer, quantity, and estimated life of the shingling. Try to find out how much of the estimate is for materials and how much for labor. He may be reluctant to tell you, but ask anyway. Check out the cost of the same materials at the local building supply house. The contractor should, of course, pay less than the retail price—since he will buy at wholesale—but at least you will have a rough idea as to what you can save by doing it yourself.

Assuming all else to be equal, it is logical to accept the contract with the lowest bid. Since all is not always equal, however, make sure that the lowest bidder is providing the same (or better) quality materials as the others. Brands may be different, but quality between the same grades (see Chapter 4) should be approximately the same. Be sure to call the references of the lowest bidder, and check such points as promised completion dates, insurance certificates, and any other differences or extras among the bidders. You may sometimes find that one of the higher bids is a better choice.

The contract. Completion dates, materials, insurance, responsibilities for cleanup of debris—all of these should be in the contract. If you want to do a portion of the work yourself, and have a contractor do the balance, make sure that it is clearly spelled out who does what, and for how much. Many contractors will not go along with such an arrangement, so make sure before you contract with him that you are in agreement.

Insurance, Bonds and Licenses

Insurance. Whenever you hire someone to work on your roof (or any other part of the house), ask the contractor to show you his Certificate of Insurance. The contractor should have both liability (for injury to others) and workers' compensation (for injury to his employees) insurance. If not, don't hire him. If you do not have a comprehensive homeowners' insurance policy of your own (and you should) take one out now to provide double protection. With both you and the contractor insured, you don't have to worry about the legal consequences of any accidents.

Licenses. Some municipalities require that contractors be licensed. Check the local building or license department to see if such protection is required. If it is, make sure that you deal only with a licensed contractor. If not, a certain amount of security is lost, but you can't expect a contractor to get a license if none is available. A license is no guarantee of superior work in any case, but it does mean that the contractor meets minimum standards.

Performance "bonds." Bonding is not generally applicable to home roofing work. A large contractor may be bonded, but most residential contractors are too small to qualify for performance bonds. If you hire a contractor who also does commercial work, it is wise to ask if he carries a performance bond that may apply to your house. If so, take down the information. If he then does not finish the job, or performs shoddy work, you may qualify for relief from the bonding company.

Financial guarantees. The best insurance is to withhold your final payment to any contractor until the work has been completed to your satisfaction. Hold out a certain amount (25 to 33⅓%) until the job is properly finished, and do not give money up front (before he starts the work) unless you cannot find a contractor who will do it otherwise.

Three-day "Change-of-Mind" law. If, in a weak moment, you sign any home-improvement or other contract that you regret as soon as the salesman leaves (or even before), you do have an escape. Although it is much better not to sign at all, you have three days, under federal

Wooden shingles can surface side walls as well as roofs, as shown. Shakes are used on the roof.

Unusually shaped shingles such as these are no longer used. These shingles are more difficult to prepare for complete reroofing than are the 3-tab.

Large individual asphalt shingles are not seen frequently anymore.

legislation, to renege on the contract. You must notify the contractor, in writing, that you do not wish to go through with the contract. Contractors, especially dishonest ones, will try to convince you that your disavowal isn't legal, but it is, and you should win if they take you to court. Legal advice is advisable if the contractor persists in trying to hold you to the contract.

FINANCING

Unless you are better off than most of us, a major roofing job will involve a loan of some sort. There are quite a few ways to get the money, but that depends on your credit rating, how "borrowed out" you are, and the current availability and cost of credit. After you scan the various methods discussed below, you may decide to do only essential repairs right now, and put off the more major items in the hope that the financial picture will improve. With rampant inflation, however, such a decision may only increase the long-run cost, because materials prices will increase each year.

Types of Loans

You will have to check with local banks for current interest rates, availability, and applicability of the terms and types listed below.

HUD (Department of Housing and Urban Development) loans. HUD is now the administrator of the old FHA (Federal Housing Authority) loans. Many banks, contractors, realtors and others still call them "FHA" loans. "Long-term" loans are designed for structural changes such as room additions. If you are adding a dormer or altering the structure of your home in addition to putting on a roof, check into this type of loan.

Most HUD loans come under "Title I" of the legislation. These loans stipulate that the home improvement must "enhance the basic livability" of the home. Any of the projects in this book should qualify under that heading. The loans are made through regular lending institutions and are insured by the federal government. The limit as of this writing is $10,000, with up to seven years to pay. Interest rates and availability vary with the money supply, but you will ordinarily pay somewhere between the prevailing mortgage rate and the cost of a standard loan. Insurance costs add another ½% to the cost.

Standard bank home improvement loan. At one time, this was the most expensive type of loan, but high rates, plus the insurance, have made some standard loans more attractive than formerly. Usually, you will need a better credit rating than for a government-sponsored loan.

Life insurance loans. If you have an endowment, whole-life or other insurance policy that has a cash value, you can borrow up to 95% of whatever the current value is. Rates are very low; contact your insurance agent as to exact rates, current cash value and other particulars.

Credit unions. Most credit unions operate through employers, fraternal or similar groups. If the credit union is with your employer, the loan is deducted from your paycheck; the lender can attach your future checks, profit-sharing accounts, or similar cash if you leave. Therefore, even those with poor credit usually can get a loan through a credit union. Rates are reasonable, but in times of high interest rates, the loans may dry up due to lack of deposits. Smaller, newer credit unions may have a limit on the amount you can borrow.

Open-end mortgages. This type of mortgage was once very common, but banks have stopped issuing them because you could borrow up to the principal amount of an older mortgage at the original interest rate. At one time this was easy to do, but lending institutions are now reluctant to abide by these terms.

Refinanced mortgages. Lending institutions are more prone to refinancing old mortgages than they used to be. You need not go to the same bank that has your mortgage, but you may have better success there. Ordinarily, this is a poor financial deal, because the entire mortgage is refinanced at a higher rate. However, it might be your only choice.

Secondary mortgage loans. These are available in most states, and are another method of borrowing on the increased equity in your home. Interest rates are relatively high, but better than a standard bank loan. Lenders usually are less tolerant of skipped payments, however, than are the primary mortgage holders—and are more prone to cause foreclosure. Be sure you can make the payments on time before you risk losing your home.

4

ROOFING MATERIALS

RECOGNIZING ROOFING MATERIALS

Before you can make even a minor roof repair, and certainly before you consider reroofing, you must be able to recognize the type of roofing material on your home. As mentioned previously, the type of roof you have, or will have, depends very much on where you live. The majority of homes in temperate climates will have asphalt shingles, which are easy enough to recognize. To test the shingle, go up a ladder to the roof or reach out a window and lift a shingle. If it bends, it is asphalt.

Wood shingles or shakes are impossible to disguise. A detailed description is given later in this chapter, but even if you do not know at this point whether they are shingles or shakes, you can tell whether they are wood.

Other possibilities for those in temperate zones are cement asbestos, slate and—more rarely—roll roofing or clay tile. Most are easy to identify. You could be fooled, however, by cement asbestos. These may look like asphalt or slate from afar. But if you try to bend one you will find that it is stiff, not flexible. Also, asbestos tiles are usually pale reddish-brown in color, and are much thinner than slate or clay tile.

Although flat clay tiles exist, by far the most common form is the semi-rounded type universally associated with Mediterranean or Spanish-style homes. Don't assume it's clay until you take a closer look, since many companies now make metal tiles that resemble clay tiles. These are quite popular because of their lower price. Metal "clay" tiles are thinner. To test, touch one.

If your roof is flat or very low in pitch, your roof probably is of built-up or roll roofing. It is common practice in the Southwest to cover the asphalt roofing with a layer of stone or gravel. The stone merely acts as a shield against the hot rays of the sun, much as the mineral granules in shingles do. Underneath it all are built-up layers of roofing felt. Roll roofing is uncommon in residential use, but is easily recognizable by the long flat sheets, its regular joints and untextured surface.

ESTIMATING ROOFING NEEDS

If you have determined that all or a substantial portion of your roof needs replacement, your next steps are deciding what type and how many roofing materials you must buy.

Direct Measurement

The simplest and most effective method of estimating roofing needs is to physically measure the area to be covered. If access is not too difficult, a folding or pull-out rule can accurately determine the dimensions of each of the roof planes. A long pull-out rule is best because you can hook it over the edge and stretch it out to its maximum length.

With multiple planes, it is best to make some rough drawings of each plane, and mark the measurements as you go along. Multiply the length and width of each plane and add them.

Figuring the "squares" needed. Once you have the square footage of the entire roofing area, divide by 100 to find out how many "squares" you need. Roofing is sold that way, by squares, with each square covering 100 square feet.

Final quantities. If you want to be precise, subtract gaps in the roofing for chimneys, skylights, or whatever. But, unless these gaps are large, it won't

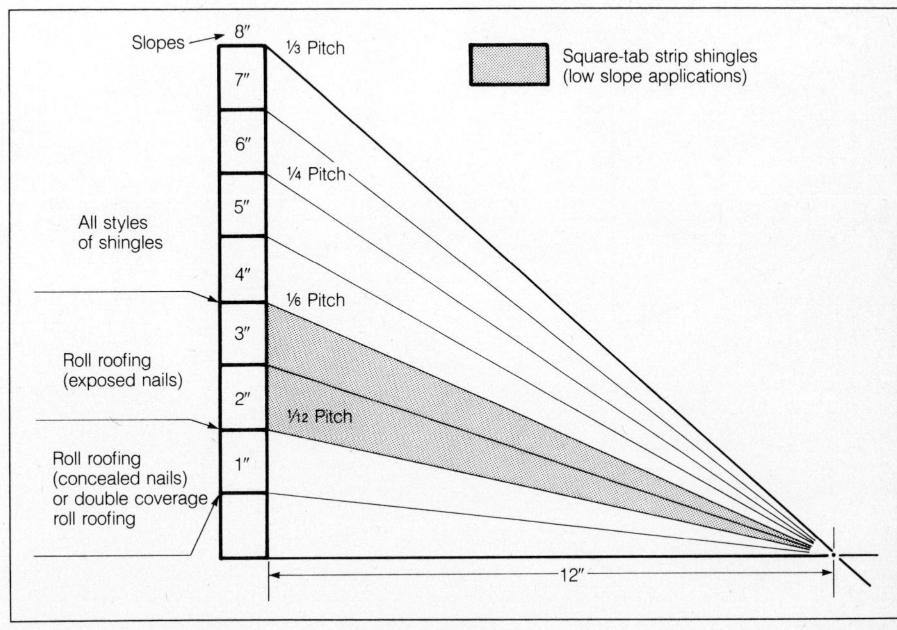

The type of shingle or roofing that is suitable will depend partially upon the pitch of the roof. Special procedures are required for low slope and steep roofs.

matter much. You should add 10% for waste, anyway, and most dealers will accept returns of full squares for credit. It is better to order too much than too little, because of slight color gradations among different lots. If you run out of materials before the job is done, and have to purchase an extra square or two, any gradation in color between batches may make the additional portion stand out.

Here's an example of how to estimate roofing needs by direct measurement of the roof shown below at right.

Roof with hips and valleys. Although hip roofs and planes with valleys may seem hard to estimate by direct measurement, these roofs are not difficult if you reduce them to planes, and remember a simple mathematical equation: The area of any triangle is half the area of the width times the height. The height is determined by drawing a line from the apex (peak) to the base.

ROOFING MATERIAL ESTIMATE

Roof	Measurement	Area
Plane 1	20x32	640 sq. ft.
Plane 2	20x32	640 sq. ft.
Plane 3	12x16	192 sq. ft.
Plane 4	12x16	192 sq. ft.
Total		1664 sq. ft.
Waste (10%)		167 sq. ft.
No. of Squares		1831 sq. ft.
Order		19 squares

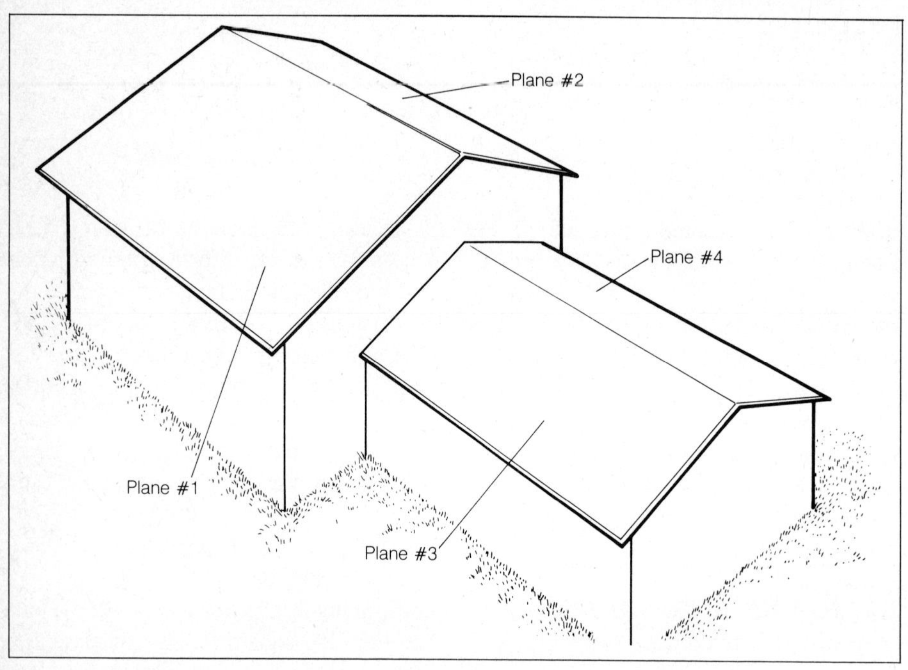

When measuring the roof in order to estimate the materials needed, first divide the roof into planes and then add up the squares needed for each plane.

Note that it isn't necessary to know the distance of all sides of the triangle, only the base. To find the area of any triangular roof plane, measure the bottom distance of the triangular area, then measure from the top straight down to the base, at a right angle. You may not be positive that you have an exact 90-degree angle at the base, but you should be able to approximate this pretty well. If you are not sure, remember that a true right angle will be the shortest distance between the base and the apex. Do not forget to halve the area after you multiply the width times the height.

If you use these mathematical principles, and study the examples given, you should be close enough so that errors will be minimal and make little difference in

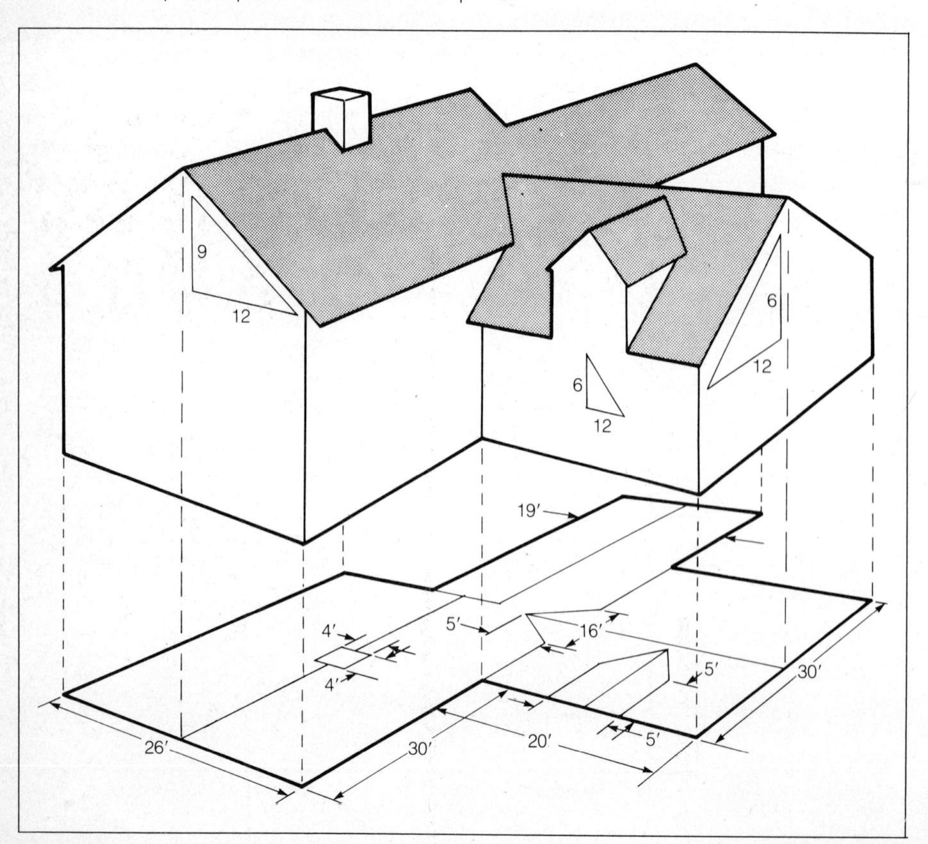

To measure a complicated roof for material estimating purposes, first reduce the various surfaces involved to a ground plan and draw up a working guide.

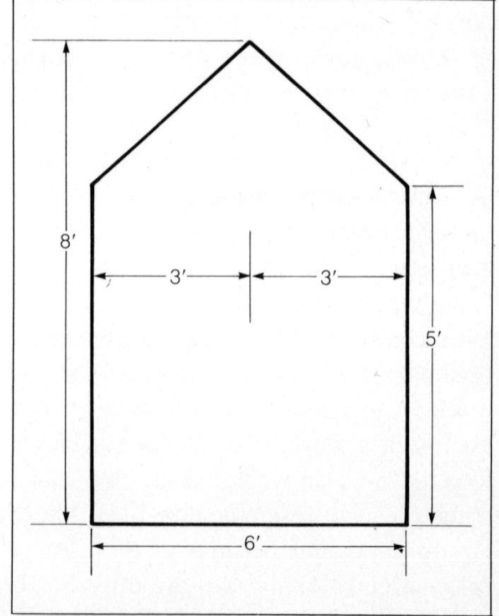

For a complicated-looking roof, divide it into squares and triangles. Then calculate the needs for each square and triangle.

the overall amount. You should be able to return any full shingle squares.

Measuring From the Ground

If you can't or don't want to go up on the roof, a contractor will probably be hired because if you can't get up there to measure, you certainly won't be doing the work. Since contractors are not always known for their honesty, you may want to check prices and quantities needed. You also will get a better idea of what's available if you do some preliminary shopping on your own.

There is a way to determine exact roofing needs from a safe perch on the ground. However, in most cases the roof area will match the perimeter of the house on the ground. The perimeter of the roof can be determined with a fair degree of accuracy by making measurements of the outside of the house. The problem lies in converting that into roof area, which will always be larger when there is any pitch to the roof. (A flat roof would be exactly the same as the perimeter.)

Using a folding rule. You can find the pitch by using a folding rule as a sort of sextant. Stand far away from the gable end of the house across the road—and hold your arms out full length. Fold the rule so that the first folding joint forms a peak. If your rule folds every six inches, which is standard, create a triangle whose slopes will be indicated by the 1-to-6 inch section and the 6-to-12 inches section. Unfold another two sections and complete the triangle. Line up the length of the rule—from the 12-inch mark to the 24-inch mark—so the triangle base aligns with a definite horizontal line of your house. Keeping the bottom horizontal alignment steady, adjust the two sections of the ruler at the joint until the two sections exactly follow the line of the roof slopes.

The beginning section of the rule will cross the horizontal section at a point between the 20 and 24 inch mark. Note the point at which the center of this end crosses the middle of a measurement mark, as shown. Read the measurement from the exact center of this intersection of the horizontal and the angle. Refer to the chart to determine the pitch/rise of your roof. If the rise is steeper than 10 in 12, accept the fact that this is a job for a professional.

Finding the pitch. In the illustrations,

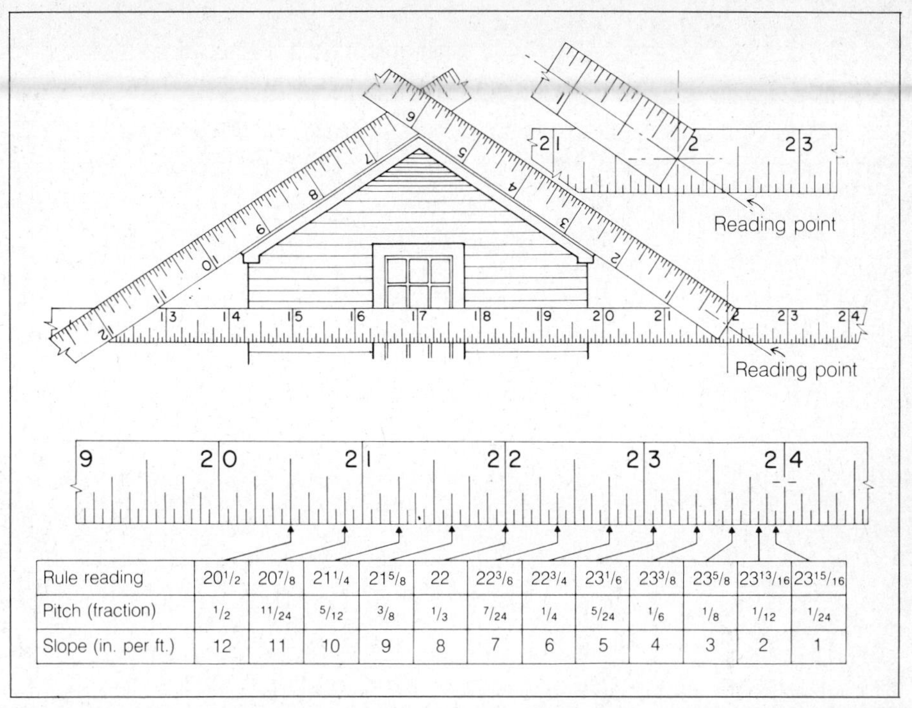

Rule reading	20 1/2	20 7/8	21 1/4	21 5/8	22	22 3/8	22 3/4	23 1/6	23 3/8	23 5/8	23 13/16	23 15/16
Pitch (fraction)	1/2	11/24	5/12	3/8	1/3	7/24	1/4	5/24	1/6	1/8	1/12	1/24
Slope (in. per ft.)	12	11	10	9	8	7	6	5	4	3	2	1

To determine roof pitch while still on the ground, use a folding rule as a sextant. Line up the bottom with the siding and the two top sections with the roof. Translate the reading point into the pitch by following the arrows to the figures below.

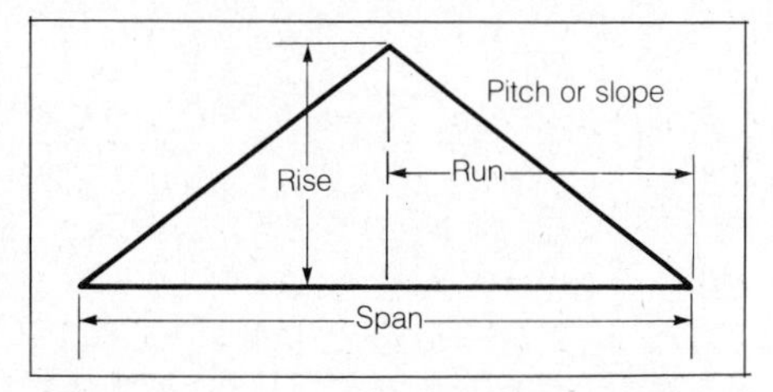

Pitch is the ratio of the rise of the roof to the span of the roof. Slope is the ratio of rise in inches to horizontal run in feet (run equals half the span).

the reading point is exactly 22 inches on the rule. Using the chart, this converts to a pitch of 1/3, or a rise (in more common roofing terms) of eight inches per foot. Any pitch can be determined by the same method. If, for example, the reading point is 21 5/8 inches on the folding rule, the pitch is 3/8, or a rise of nine inches per foot. A reading of 23 3/8 would mean a pitch of 1/6, a rise of four inches per foot.

Final Quantities

When the pitch and the horizontal square footage are known, the total amount of roofing can be determined using the conversion factors shown in Table 4.1. For example, with a horizontal square footage of 1000, and a pitch of four inches per foot (1/6 slope), the amount of roofing needed will be 1054 sq. ft. If the area is more than 1000 sq. ft. (for example), add the amount for both 1000 and 200 square feet (1054 plus 210.8): a total of 1264 sq. ft. If you add 10 percent for waste—126 square feet in the last example—you would order 1390.8 square feet of roofing materials, or 14 squares.

Estimating Other Materials

The calculations above tell you how much "field" shingling you will need for the roof deck. There are other materials, however, which you also will need to estimate. Starter strips are ordinarily used along the eaves, and these can be measured horizontally. The same applies for edging strips along eaves. Ridge shingles can be cut from asphalt strip shingles, but you can order special ridge shingles if you prefer. For wood, and most other types of shingles, special ridge shingles are highly recommended. Unless you have a hip roof (in which case, see below), the ridge can also be measured on the horizontal.

Edging strips for rakes. Estimate by using the same conversion table for pitches in Table 4.1. If the horizontal rake measurement is 30 feet, for example, with a pitch of four inches per foot, you will need 31.6 feet of materials for that rake.

Hip and valley measurement. These are a little more complicated. Use the horizontal length under the longest rafters and divide it in half. This gives the "run of the common rafters". The first column of Table 4.2 is the run of the common rafter in feet. If that figure is 10 feet, for

TABLE 4.1 RISE AND PITCH CONVERSIONS*

Rise (Inches per ft. of horizontal run.)	1″	2″	3″	4″	5″	6″	7″	8″	9″	10″	11″	12″
Pitch (Fractions)........	1/24	1/12	1/8	1/6	5/24	1/4	7/24	1/3	3/8	5/12	11/24	1/2
Conversion Factor	1.004	1.014	1.031	1.054	1.083	1.118	1.157	1.202	1.250	1.302	1.356	1.414
Horizontal (Area in Sq. Ft. or Length in Feet)												
1	1.0	1.0	1.0	1.1	1.1	1.1	1.2	1.2	1.3	1.3	1.4	1.4
2	2.0	2.0	2.1	2.1	2.2	2.2	3.2	2.4	2.5	2.6	2.7	2.8
3	3.0	3.0	3.1	3.2	3.2	3.2	3.5	3.6	3.8	3.9	4.1	4.2
4	4.0	4.1	4.1	4.2	4.3	4.5	4.6	4.8	5.0	5.2	5.4	5.7
5	5.0	5.1	5.2	5.3	5.4	5.6	5.8	6.0	6.3	6.5	6.8	7.1
6	6.0	6.1	6.2	6.3	6.5	6.7	6.9	7.2	7.5	7.8	8.1	8.5
7	7.0	7.1	7.2	7.4	7.6	7.8	8.1	8.4	8.8	9.1	9.5	9.9
8	8.0	8.1	8.3	8.4	8.7	8.9	9.3	9.6	10.0	10.4	10.8	11.3
9	9.0	9.1	9.3	9.5	9.7	10.1	10.4	10.8	11.3	11.7	12.2	12.7
10	10.0	10.1	10.3	10.5	10.8	11.2	11.6	12.0	12.5	13.0	13.6	14.1
20	20.1	20.3	20.6	21.1	21.7	22.4	23.1	24.0	25.0	26.0	27.1	28.3
30	30.1	30.4	31.0	31.6	32.5	33.5	34.7	36.1	37.5	39.1	40.7	42.4
40	40.2	40.6	41.2	42.2	43.3	44.7	46.3	48.1	50.0	52.1	54.2	56.6
50	50.2	50.7	51.6	52.7	54.2	55.9	57.8	60.1	62.5	65.1	67.8	70.7
60	60.2	60.8	61.9	63.2	65.0	67.1	69.4	72.1	75.0	78.1	81.4	84.8
70	70.3	71.0	72.2	73.8	75.8	78.3	81.0	84.1	87.5	91.1	94.9	99.0
80	80.3	81.1	82.5	84.3	86.6	89.4	92.6	96.2	100.0	104.2	108.5	113.1
90	90.4	91.3	92.8	94.9	97.5	100.6	104.1	108.2	112.5	117.2	122.0	127.3
100	100.4	101.4	103.1	105.4	108.3	111.8	115.7	120.2	125.0	130.2	135.6	141.4
200	200.8	202.8	206.2	210.8	216.6	223.6	231.4	240.4	250.0	260.4	271.2	282.8
300	301.2	304.2	309.3	316.2	324.9	335.4	347.1	360.6	375.0	390.6	406.8	424.2
400	401.6	405.6	412.4	421.6	433.2	447.2	462.8	480.8	500.0	520.8	542.4	565.6
500	502.0	507.0	515.5	527.0	541.5	559.0	578.5	601.0	625.0	651.0	678.0	707.0
600	602.4	608.4	618.6	632.4	649.8	670.8	694.2	721.2	750.0	781.2	813.6	848.4
700	702.8	709.8	721.7	737.8	758.1	782.6	809.9	841.4	875.0	911.4	949.2	989.8
800	803.2	811.2	824.8	843.2	864.4	894.4	925.6	961.6	1000.0	1041.6	1084.8	1131.2
900	903.6	912.6	927.9	948.6	974.7	1006.2	1041.3	1081.8	1125.0	1171.8	1220.4	1272.6
1000	1004.0	1014.0	1031.0	1054.0	1083.0	1118.0	1157.0	1202.0	1250.0	1302.0	1356.0	1414.0

**Use for conversion of horizontal distances or areas to slope distances or areas.*

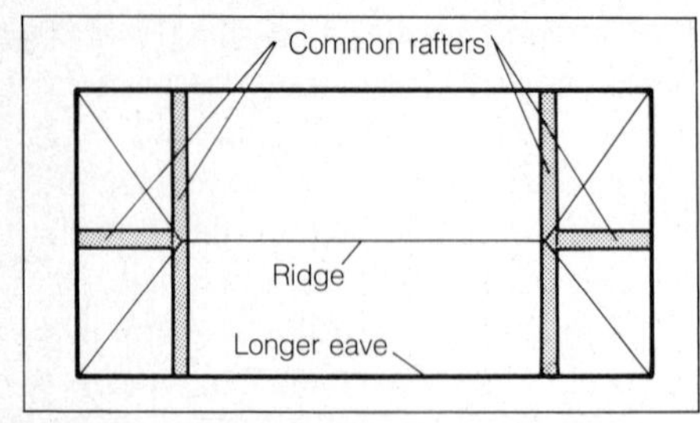

Common rafters are the longest rafters. Hip roof (shown) has a common rafter at each end and several between the ridge pole and long eaves.

example, with a pitch of six inches, the amount of flashing or other material needed for that hip or valley is 15 feet.

Doing the Measuring Yourself?

The preceding charts and figures have been given for the sake of completeness. However, there is little reason to take measurements from the ground when there is ready access to the roof—and if the roof is difficult to reach, you don't want to do the roofing work yourself. In these cases, where you will hire a contractor, leave the measuring to him as well. One such situation is a roof with curved or circular areas, a case that is admittedly rare. If you have this type of roof, see a contractor. You'll have a difficult time roofing the house even if you figure out how to measure it.

TYPES OF SHINGLES

Asphalt or Fiberglass Shingles

There are at least half a dozen manufacturers of asphalt and fiberglass shingles, and they all call their grades and styles by different names. However, they all produce basically the same materials. Choosing between manufacturers is most often picking from what is available or the best price. All brands are manufactured to rigid standards. Any roof can be covered with asphalt or fiberglass shingles, except flat roofs or those with a pitch of less than 2-in-12 (1/6).

Fire ratings. The choice between asphalt or fiberglass involves one major factor—fire resistance. All shingles of this type are reasonably fire-resistant. Most asphalt shingles carry a rating of Class C rating. Almost all communities require that roofs meet this requirement; do not buy any shingles that do not meet it. A roof with Class C shingles is very resistant to fires, but fiberglass, with a

TABLE 4.2 DETERMINING LENGTHS OF VALLEYS AND HIPS

Rise {Inches per foot of horizontal run}	4″	5″	6″	7″	8″	9″	10″	11″	12″	14″	16″	18″
Pitch {Degrees.	18° 26′	22° 37′	26° 34′	30° 16′	33° 41′	36° 52′	39° 48′	42° 31′	45°	49° 24′	53° 8′	56° 19′
Pitch {Fractions	1/6	5/24	1/4	7/24	1/3	3/8	5/12	11/24	1/2	7/12	2/3	3/4
Conversion Factor	1.452	1.474	1.500	1.524	1.564	1.600	1.642	1.684	1.732	1.814	1.944	2.062
Horizontal* (Length in Feet)												
1	1.5	1.5	1.5	1.5	1.6	1.6	1.6	1.7	1.7	1.8	1.9	2.1
2	2.9	2.9	3.0	3.0	3.1	3.2	3.3	3.4	3.5	3.6	3.9	4.1
3	4.4	4.4	4.5	4.6	4.7	4.8	4.9	5.1	5.2	5.4	5.8	6.2
4	5.8	5.9	6.0	6.1	6.3	6.4	6.6	6.7	6.9	7.3	7.8	8.2
5	7.3	7.4	7.5	7.6	7.8	8.0	8.2	8.4	8.7	9.1	9.7	10.3
6	8.7	8.8	9.0	9.1	9.4	9.6	9.9	10.1	10.4	10.9	11.7	12.4
7	10.2	10.3	10.5	10.7	10.9	11.2	11.5	11.8	12.1	12.7	13.6	14.4
8	11.6	11.8	12.0	12.2	12.5	12.8	13.1	13.5	13.9	14.5	15.6	16.5
9	13.1	13.3	13.5	13.7	14.1	14.4	14.8	15.2	15.6	16.3	17.5	18.6
10	14.5	14.7	15.0	15.2	15.6	16.0	16.4	16.8	17.3	18.1	19.4	20.6
20	29.0	29.5	30.0	30.5	31.3	32.0	32.8	33.7	34.6	36.3	38.9	41.2
30	43.6	44.2	45.0	45.7	46.9	48.0	49.3	50.5	52.0	54.4	58.3	61.9
40	58.1	59.0	60.0	61.0	62.6	64.0	65.7	67.4	69.3	72.6	77.8	82.5
50	72.6	73.7	75.0	76.2	78.2	80.0	82.1	84.2	86.6	90.7	97.2	103.1
60	87.1	88.4	90.0	91.4	93.8	96.0	98.5	101.0	103.9	108.8	116.6	123.7
70	101.6	103.2	105.0	106.7	109.5	112.0	114.9	117.9	121.2	127.0	136.1	144.3
80	116.2	117.9	120.0	121.9	125.1	128.0	131.4	134.7	138.6	145.1	155.5	165.0
90	130.7	132.7	135.0	137.2	140.8	144.0	147.8	151.6	155.9	163.3	175.0	185.6
100	145.2	147.4	150.0	152.4	156.4	160.0	164.2	168.4	173.2	181.4	194.4	206.2

**The "run of the common rafter"*

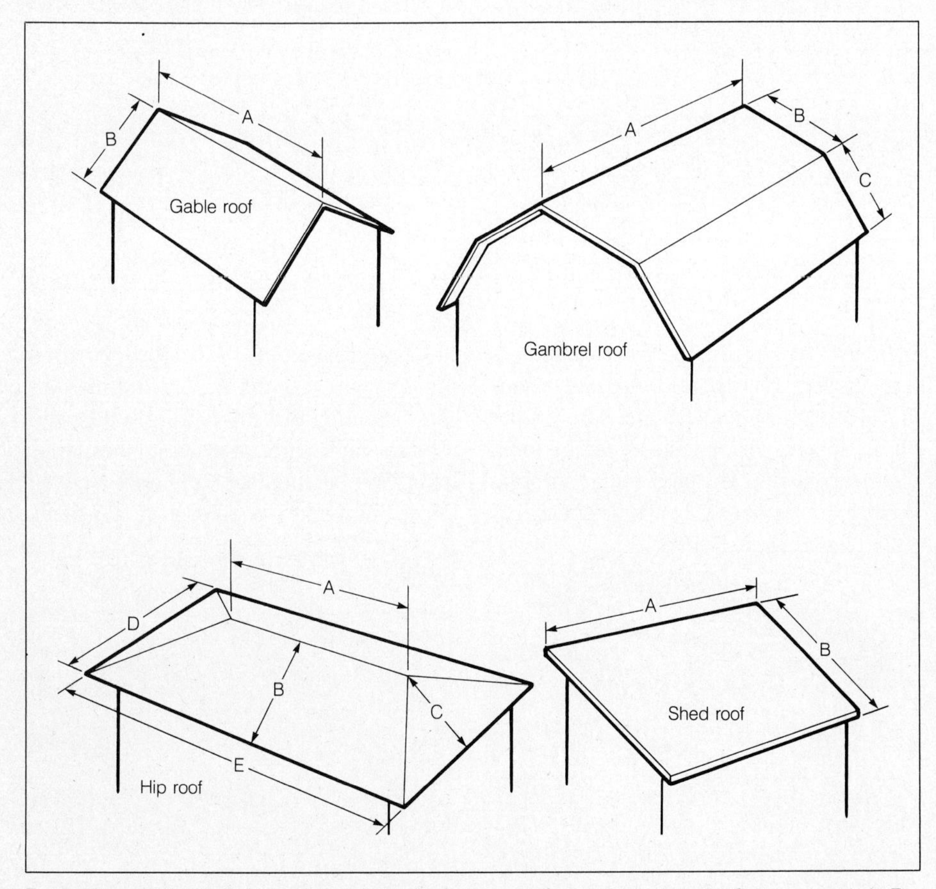

For shed, multiply A×B. For gable, double A×B. For gambrel, add A×B to A×C; then double it. For hip, add A+E; divide by 2; multiply by 2B. Add this to C×D.

Class A rating, is even better. For those in very dry climates, fiberglass shingles should be considered seriously. Class C Shingles are adequate for most areas, however, and cost quite a bit less. Look for the UL (Underwriter's Laboratory) seal on the bundles of any asphalt or fiberglass shingles that you buy. It should be Class C or better.

Asphalt shingle composition. These shingles are highly durable, and provide good weather-resistance. Although termed "asphalt", the shingles use asphalt only as a base for the mineral granules that are embedded in the asphalt layer. If only asphalt were used, there would be weather protection only for a few years, after which the asphalt would crack up and/or be worn away by the elements. Without the hard mineral granules, the shingles would be little more than a thick roll of building paper.

Size and shape. Virtually all shingles now come in strips of 12x36 inches. Most are "three-tabs" (with two cutouts), but the same-size strips are also available in two tabs or no tabs at all (straight-butt types). The strips are all laid in the same way (four nails per strip) regardless of

STRIP AND INTERLOCKING SHINGLES

Type of Shingle	Available as	Width	Length	Exposure	UL Rating
Self-sealing random tab strip shingle Multi-thickness	Assorted edge, surface texture and application treatments	11½" to 14"	36" to 40"	4" to 6"	A or C Many are wind resistant
Self-sealing random tab strip shingle Single-thickness	Assorted edge, surface texture and application treatments	12" to 13¼"	36 to 40"	5" to 5⅝"	A or C Many are wind resistant
Self-sealing square tab strip shingle	Two-tab or four-tab	12" to 13¼"	36" to 40"	5" to 5⅝"	A or C All are wind resistant
Three-tab	Three-tab	12" to 13¼"	36" to 40"	5" to 5⅝"	
Self-sealing square tab strip shingle No-cutout	Assorted edge and surface texture treatments	12" to 13¼"	36" to 40"	5" to 5⅝"	A or C All are wind resistant
Individual interlocking shingle Basic design	Design variations	18" to 22¼"	20" to 22½"		C Many are wind resistant

how many cutouts there are. Whether you prefer single-butt, or two-tab or three-tab, is a matter of taste and personal preference. Three-tab is by far the most common. Part of each tab is covered by the row of shingles above, and part is exposed to the weather. The covered dimensions is that part of the shingle ''laid to the weather''. The amount of shingle not covered is the ''exposure''.

Strip shingle tabs can be trimmed or offset in order to achieve straight or staggered butt lines. They also can be purchased embossed or built up, to offer a three-dimensional effect. Each shingle characteristic—staggered butt line, lamination, or embossing—can be combined. This will create textures to resemble wood, slate and tile.

Interlocking shingles. This design offers greater resistance to strong winds than with strip shingles. They are available in several shapes and types of locking mechanisms. The large single shingles usually are rectangular or hexagonal.

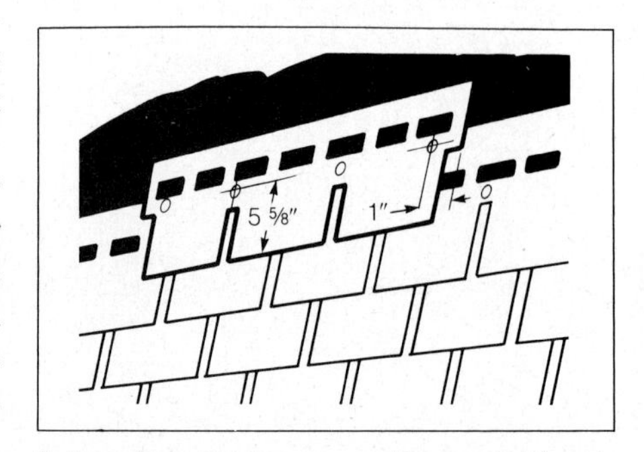

Self-sealing shingles have a thermoplastic adhesive just above the nailing line to keep shingles from blowing off in heavy winds.

Wind resistance. Most quality roofing now also carries a UL wind-resistant seal, meaning that the roofing is designed to withstand winds of at least 60 miles an hour. Self-sealing tabs were once a novelty, but now most good roofing has them.

Replacement of an old three-tab shingle with a new one usually involves just sliding in the new shingle and nailing it in place.

ROOFING STANDARDS (ASPHALT AND FIBERGLASS)

All manufacturers who are members of the ARMA (Asphalt Roofing Manufacturers Association), which includes the major manufacturers of both asphalt and fiberglass shingles, conduct inspections to assure the quality of their products. The major points of inspection are:

1. Saturation of felt to determine quantity of saturant and efficiency of saturation
2. Thickness and distribution of coating asphalt
3. Adhesion and distribution of granules
4. Weight, count, size, coloration and other characteristics of finished product before and after packaging.

UL DESIGNATIONS FOR ROOFING (ASPHALT AND FIBERGLASS)

The designation of asphalt or fiberglass roofing as Class A, B or C by the Underwriters' Laboratories, Inc., applies to the degree of fire resistance. It means that any shingle containing the UL symbol is not easily ignited, will not readily contribute to the spread of flame, and does not possess any "flying brand" hazard. The latter means that the shingle would not become a flaming or glowing torch if it falls or is blown off the deck during a fire. The differences between the letters are:

Class A Effective against *severe* fire exposure
Class B Effective against *moderate* fire exposure
Class C Effective against *light* fire exposure

Most shingles manufactured today also carry the UL designation for wind resistance. Such shingles have either factory-applied adhesive on the tabs, which melt with exposure to the sun, or they interlock. These symbols are on the bundles or packaging rather than on the shingles.

These contain a factory-applied "thermoplastic" adhesive on the bottoms of the tabs, which is dry when laid on the roof. The heat of the sun melts the adhesive and bonds each shingle to the one below. Previously, the extra cost of the self-sealant made them practical only for high-wind areas, but now the extra cost is minimal. Self-sealing tabs are well worth the small extra cost and are highly recommended for all areas.

Color. In hot and humid areas of the southern U.S., white shingles are most popular because they provide the best reflectance against the rays of the sun. But white shingles are most susceptible to staining and to discoloration resulting from fungi and algae in these regions. Other colors are affected, but the discoloration is less noticeable. If you live along the Gulf of Mexico, or other hot, humid regions, South Atlantic Coast, try to find shingles that are pretreated to withstand algae and fungi. This is not a factor in most other areas. Black is the most practical color for cold regions.

Life expectancy, thickness and weight. Manufacturers at one time made somewhat exaggerated claims as to life expectancy of their products, a practice that has been discouraged by the Magnuson-Moss Warranty Act of 1975. Former life-expectancy figures are now put in terms of limited warranties. The limited warranty figures are now somewhat modest. If a manufacturer will extend a limited warranty for 15 years, you should be able to get 20 years of life out of the roof under average weather conditions.

Where do the warranty and life-expectancy figures come from? Primarily, from the weight and thickness of the shingles. In other words, the heavier and thicker the shingle, the longer it can be expected to hold up. At one time, the heaviest shingle weights were 235 pounds per square (hundred sq. ft.). Modern manufacturing methods, such as laminating two layers of asphalt together, have enabled weights of 350 pounds and more per square. Double-layered and regular heavy-weight shingles are usually warranted for 25 years or more.

Fiberglass vs. asphalt. Although similar to asphalt in exterior appearance, fiberglass shingles are quite different in terms of weight and life expectancy. Longer life expectancy can be anticipated from fiberglass shingles although they are thinner and lighter than comparable asphalt shingles. That is because the asphalt-saturated mat is replaced by a stronger backing of glass fibers, which gives at least as good a weather resistance as most heavier and thicker shingles.

Purchasing hints. Since application costs (or time, if you are doing it yourself) are about the same no matter what type of shingle you use, it makes sense to buy the heaviest, best shingles you can afford. If you are doing it yourself, remember that this is a time-consuming job, which you won't want to repeat too soon. Figure that the heaviest shingles cost about twice as much as the cheapest ones.

There is one case, however, in which you might not want to put up heavy shingles. It is possible that the extra weight of heavy shingles over old ones might be too much for the structure to support. When applying heavy shingles over old, check that the roof can support the load.

Wood Shingles and Shakes

At one time, cedar shingles or shakes were widely used for roofing because of their inexpensive cost. Times have changed, however, and wood shingles—

Wood shingles, replaced individually for minor roof repairs, will look conspicuous until they have weathered, but the difference will be less noticeable from the ground.

almost exclusively made of western red cedar—now cost considerably more than asphalt shingles. They are also more difficult to apply and have poor fire resistance.

The overwhelming benefit offered by wood roofing is its appearance. Another advantage is its superior R-factor (thermal resistance). Cedar shingles ½-inch thick have an R-factor twice that of typical asphalt shingle (.94 as compared to .44 for asphalt shingling). Heavy, hand-split shakes with a one-inch average butt thickness have an R-factor of 1.69. This compares well with an R-factor of 05 for ½-inch slate, or asbestos-cement shingles with .21.

If you can afford wood shingles, which cost about twice as much as the heaviest asphalt shingles, buy the top grade, called #1 Blue Label. These are 100 percent clear edge-grain heartwood.

Shingles vs. Shakes

Shingles are cut by machine. Shakes are split by hand on at least one face, and have a rough, rustic look that cannot be duplicated by machine.

Both wood shingles and shakes come packaged the same as for asphalt shingles. Each square is designed for ordinary applications of 100 square feet. There are four bundles in each square. Wood shingles come in several different sizes, with varying exposures. Each bundle covers 25 square feet.

Other Choices

There are choices other than asphalt or wood shingles for roofing materials, but it's doubtful that anyone would buy them for do-it-yourself work. Some of them are still seen on older homes, and you may need a few for repairs.

Clay tile. Clay tiles and their metal lookalikes are popular on Spanish-style homes, particularly in the U.S. Southwest. Again, however, it is unlikely that anyone would reroof his own house with clay tile. That's a job for a specialist. Furthermore, any house that is not already roofed with clay tile would probably look odd with a new clay roof. If you now have clay tile, it should last the life of the house.

Cement-asbestos. This is a fireproof tile that is durable but heavy, expensive, and very difficult to find and apply. A special cutter is needed to cut through the tiles; the equipment can be rented. Since asbestos is a major component of these shingles, we do not recommend cutting them yourself. In addition to the cutting difficulty, asbestos is a dangerous carcinogen. Asbestos fibers have a virulent effect on the lungs, causing a deadly form of cancer called mesothelioma. If you must add this type of roof, hire a contractor.

Slate. Along with many of the same advantages and drawbacks as cement-asbestos tiles, slate is quite expensive and very heavy. Ask yourself: if your roof framing can support it, how will you get the slate up there? It offers a beautiful—nearly eternal—roof, but application is a monumental, painstaking job. If you can afford slate, you can afford a contractor to lay it.

Built-up roofing. This provides a serviceable roof, but not a particularly attractive one. It should be considered only on flat or very low-slope roofs. Since it requires application of hot asphalt, hire a reliable contractor.

Metal roofing. There is the aforementioned metal "clay" tile, which is a good compromise if you want a clay tile look that is less expensive and lighter in weight. The galvanized steel base is covered with mineral granules. These come in panels of four tiles, with a choice of nine colors, weighing only 170 pounds per square. One prominent manufacturer of this product is Automated Building Components, Inc.

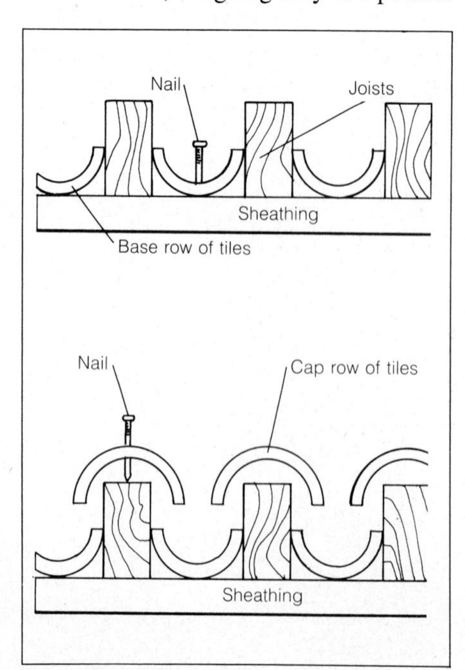

Roofs with curved tiles are actually made up of half-tiles that overlap each other. The base tiles are nailed to the roof sheathing; the cap tiles are nailed to the furring strips. For any extensive amount of repairs, call a professional.

At least one manufacturer makes aluminum "shakes" directed toward the do-it-yourself market: Reinke Mfg. Co. shakes weigh only 46 pounds per square and carry a limited warranty of 50 years.

Aluminum "shingle-shakes" are made by most of the major aluminum manufacturers such as Alcoa, Kaiser and Reynolds. These are embossed, preformed panels three to five feet long. They have an irregular surface somewhat like wood shingles, although the colors are different. Interlocking edges make them almost impervious to up to hurricane-force winds.

Concrete tile roofs. Extruded concrete tiles are sometimes used instead of clay tiles, which they resemble. They are cheaper and enable faster installation, but this still is not a job for most do-it-yourselfers—no roofing is heavier than concrete tiles, which weigh 840 to 900 pounds per square. Longevity is excellent, with warranties of up to 50 years.

AUXILIARY MATERIALS

When estimating roofing materials, don't forget the products that accompany shingle installation. These include fasteners, specialized shingles, underlayment and a host of other items.

For Shingle Application

Ridge and hip shingles. These can be cut from regular shingles, but preformed units are easier to apply, especially with wood roofing. Measure along the ridge and/or hip to estimate.

Felt underlayment. When used under asphalt or fiberglass shingles, plan on four standard rolls of 15-pound asphalt-saturated felt for every square of shingles. Standard rolls are 36 inches wide and 144 feet long. Approximately 1½ rolls of felt are used per square for shakes at 10-inch exposure. Shakes require 18-inch rolls instead of 36-inch, so half the number of rolls are needed for each square if you must cut 36-inch rolls in half.

Nails. Hot-dipped galvanized nails are essential to all types of roofing. Where possible, use roofing nails, which have flatter, oversized heads. Nails need to be long enough to go through the shingles and then penetrate the roof deck by ¾ inch. Longer nails are needed for reroof-

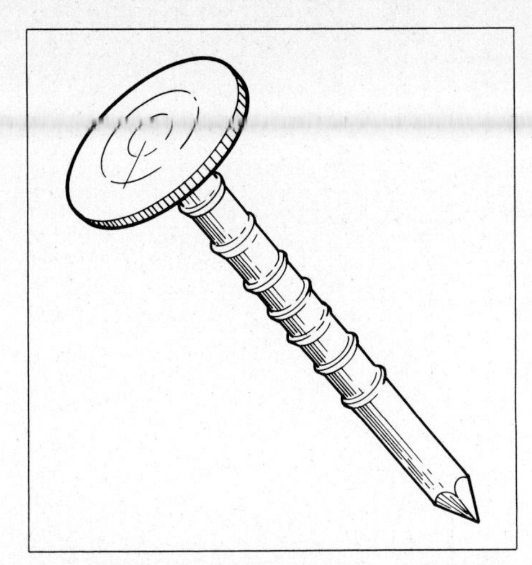

The standard roofing nails have wide heads and are slightly barbed for better holding power than is offered by regular nails.

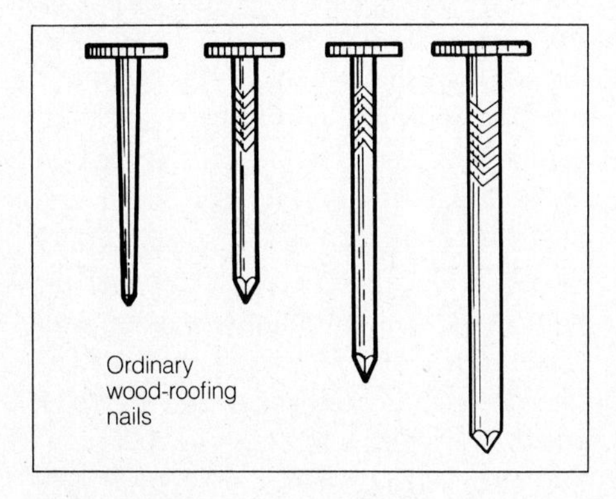

ing than for new construction or when the old roofing has been removed.

New roofing over old. To reroof with new asphalt shingles over old asphalt shingles, use nails that are 1¾ inches long (or 5d). If the existing roofing is of 24-inch shingles, use 2 inch (6d) nails. To reroof with asphalt shingles over old wood shingles, choose 1¾ inch nails. When putting new wood shingles or shakes over old, use 2½ inch (8d) nails. Choose this same length for shake hips and ridges.

Asphalt shingles, new construction, old roofing removed. You will need about 2½ pounds of 1¼ inch (or 3d) long roofing nails for each square of asphalt 16- and 18-inch shingles, and a few pounds of one-inch nails for the underlayment. For 24-inch shingles, use slightly longer (½ inch or 4d) nails. You need four nails per shingle.

Wood shingles or shakes, new construction, old roofing removed. For this construction, a little more than two pounds of box (small-head) nails per square (two per shingle). The lengths of these nails depends upon the thicknesses of the new shingles. For handsplit shakes, 2-inch (6d) nails are adequate unless the shakes are quite thick—then use 2¼ inch (7d) nails. Shake hips and ridges require 2½ inch (8d) nails.

Valley materials. For closed valleys, use one layer of 36-inch-wide 90-pound mineral-surfaced roll roofing. Open valleys require either two layers of roll roofing (one 18-inch wide and one 36-inch) or copper or aluminum flashing. Follow manufacturer's directions. Wood roofing requires a 20-inch center-crimped metal valley. Measure the length of each valley manually, or estimate.

Roofing cement. A one-gallon can should be enough for a small roof. Buy two gallons for larger roofs.

Preservatives. None usually is needed, even for wood shingles and shakes. You may want to use a wood preservative such as "penta" on wood roofs in certain areas. Observe all safety precautions.

Related Materials

Roofing and gutter tape. Use these to repair gutters, flashings and almost anything else on the roof, as long as the defect is minor. This aluminized tape is strong and durable, and is used like Scotch tape to cover and waterproof leaking, damaged or cracked areas almost anywhere. Simply clean the offending area thoroughly and press the tape over the defect.

Drip edges. Use four-inch metal or vinyl strips along eaves and rakes. Measure the lineal footage along the edges to find the amounts needed.

Flashing metal. Locations and types of flashing needs are covered in Chapters 6 and 9. Direct measurement is the only accurate way to determine amounts needed. Copper or aluminum flashings are preferred.

Sheathing. Sheathing, as discussed in Chapter 1, can be of Exterior plywood or of tongue-and-groove boards. See Chapter 1 for an in-depth discussion.

Lumber for framing. When replacing roof framing, use the same size as was previously used. Choose Structural grade No. 2 lumber to ensure adequate and longlasting strength.

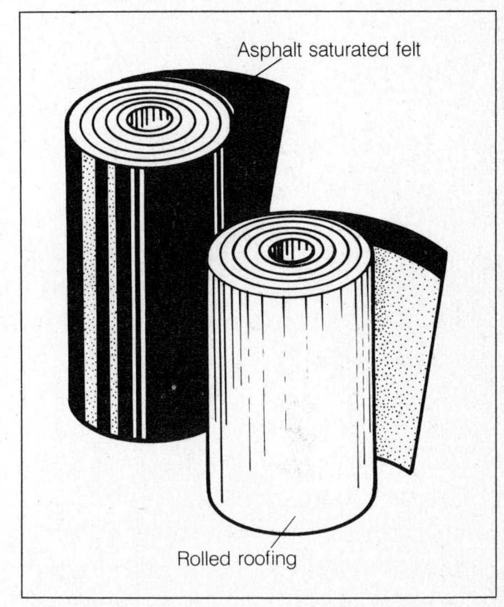

Asphalt-saturated felt is used as underlayment. Roll roofing has granular material embedded in one surface.

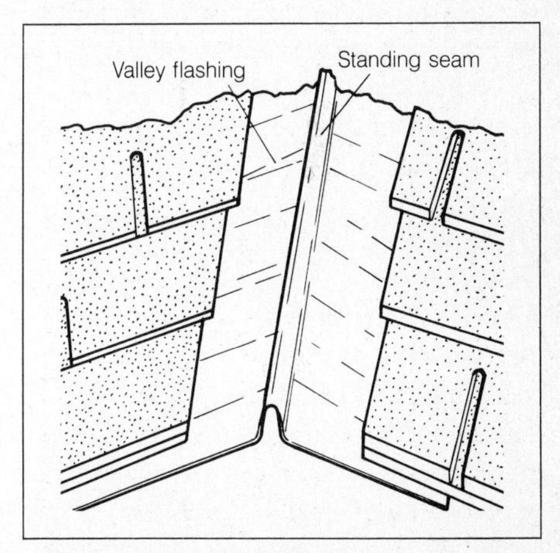

Most metal valley flashing has a "standing seam" or crimp down the center. The shingles are cut back to frame the seam.

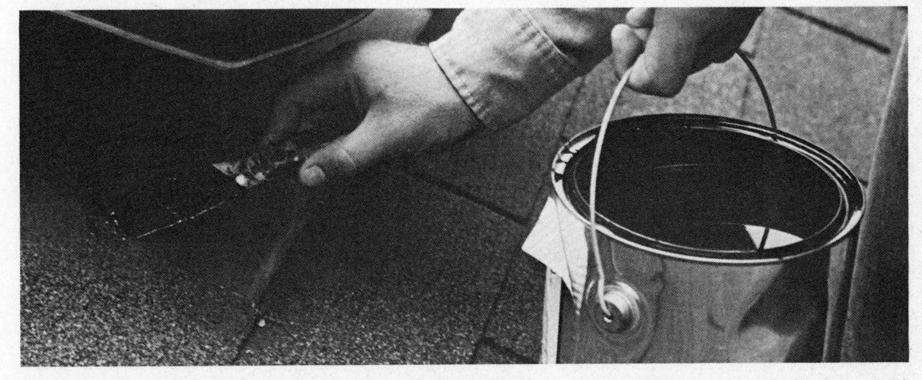

Roofing cement is a thin; liquid asphalt material. It is used to adhere loose shingles, to seal flashing, shingles, and small holes in roofing. Apply with a putty knife.

5

TOOLS AND SAFETY

Proper tools are essential in order to work safely on the roof. Most do-it-yourself jobs involve a small amount of risk. The big difference is that roofing, like electrical work, is inherently hazardous. The big danger in roofing, of course, is falling. The higher and steeper the roof, the greater the hazard. But proper tools (and common sense) can reduce the risk. Appropriate ladders and other safety tools, and their correct use, make the job both safer and easier.

Therefore, ladders and other safety devices are categorized here as roofing tools. Along with the discussion of basic tools and safety "tools", we will discuss some mandatory safety rules and suggest ways to keep your roofing work on a healthy track.

TOOLS

The Basic Tool Kit

For any type of roofing job, certain tools are essential. You just can't work without them. Fortunately, roofing requires only a few basic tools, most of which you probably own already. A standard 16-oz. claw hammer, for example, is the principal tool used for asphalt shingling.

Claw hammer. The standard 16-oz. hammer has a slightly rounded head and two curved claws for pulling out nails at the other end. Buy a good one, with a head that is firmly attached to the handle. The best ones have a metal handle padded with rubber or leather, and are forged in one piece with the head. Almost as good are those with wooden handles where the shank is held in place with triangular cleats and covered over with solidified liquid plastic.

Measuring devices. A six-foot wooden folding rule is easiest to work with on a roof. For measurement of longer distances, metal tape measures eight feet or longer are convenient. Use both, if you can.

Trowel. The type is not too important, but it is easiest to work with a pointed mason's trowel, which is used more often for brickwork than for roofing. Trowels are helpful when applying roofing cement, but a putty knife or similar tool will serve nearly as well.

Soft-soled shoes. Rubber, crepe or similar good-gripping soles are essential. Sneakers or tennis shoes are acceptable. Soft-leather work shoes are even better because they give ankle support.

Nylon rope. There are many uses for nylon rope, including creating safety lines or securing ladders. To make a safety line, nail or securely clamp a 2x4 to the inside of a window and on the face of the house opposite to where you will be working. Toss the rope over the roof to the working side. To avoid damage to the rope or roof, cut a two-foot piece of garden hose and slide it over the rope up to the ridge. This acts as a buffer and prevents chafing. Always use safety knots, as shown, in order to prevent the knot's tightening against you. Never leave slack in the line. It should only be long enough to let you reach the area where you will be working. When moving and leaving slack, wrap the rope around a chimney or other solid, immovable object.

Carpenter's square. Used for mea-

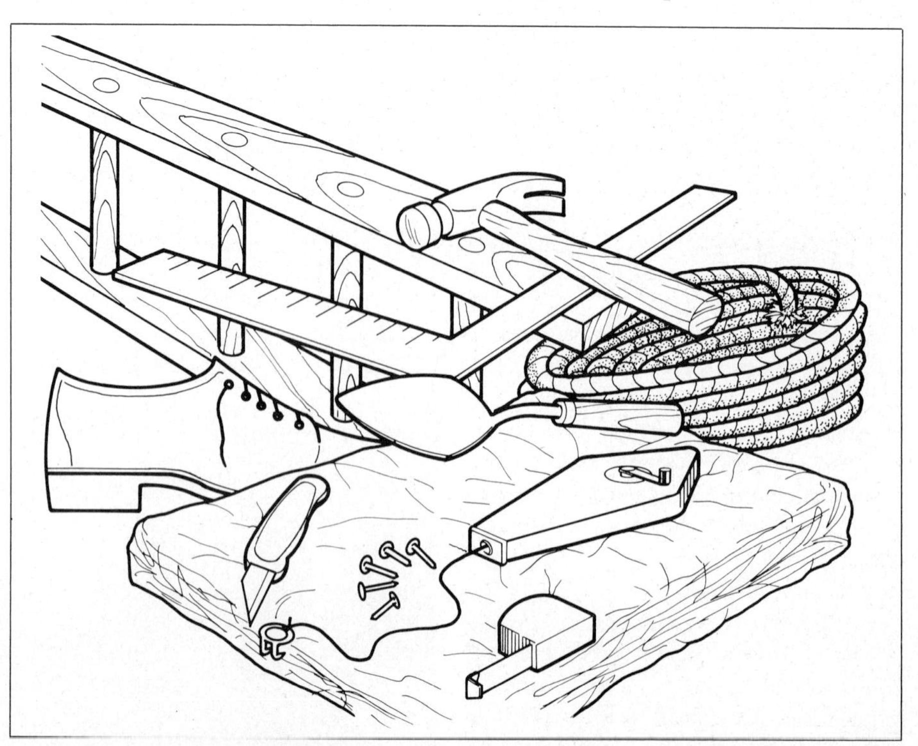

The basic tool kit (starting at top left and moving counterclockwise) includes: ladder, carpenter's square, pointed mason's trowel, rubber-soled shoes, utility knife, tarp or plastic sheet, tape measure, chalkline, nylon rope, hammer.

suring and as a straightedge for cutting shingles, its primary function is for checking alignment.

Utility knife. Also called a razor knife, this is a handy tool for cutting shingles, felt underlayment and numerous other materials. Find one with replaceable blades, although a curved "linoleum knife" can give satisfactory service.

Polyethylene sheets. For an inexpensive protection against sudden rainstorms, buy the heavy-duty large sheets used by painters as drop cloths. Keep them handy to cover up bare wood or any holes in the roof if you must leave before securing the roof against leaks. Canvas tarpaulins are better, of course, but they are much more expensive.

Chalkline and reel. This set is preferred for proper vertical and horizontal alignment in most roofing jobs, and is indispensible for valley work.

Optional Tools

You will undoubtedly need all of the preceding tools. Here are some that you may also need or find useful, depending on the type of roofing job.

Square-end shovel. What you need is the garden-spade variety, which is helpful when removing old shingles. Shovels with round or pointed tips are no help for this.

If a flat spade is not available, use an ice chipper to lift up the shingle that lies on and above the damaged one.

Crowbar. This tool can be used to remove nails and some types of shingles. A hammer serves just as well for nails, but a crowbar with a curved-end remover is easier on the back.

Pneumatic stapler. A handy tool because it makes asphalt roofing go up much faster, the stapler's drawback is that it requires a compressor; many feel the aggravation and expense are not worth it.

Nail holder. Some workers like to keep nails in pockets or a container, but a nail apron worn around the waist is best for this job.

Nailstripper. If you are putting up wood shingles or shakes, a nailstripper is a great time saver. It's not for removing nails but for lining up wood-roofing nails ready for use. Nails are loaded into the stripper, with the nail head up, so that you can grab a few at a time as you need them and without flipping them over. With one of these, you can put down four or five shingles or shakes at a time. It is worn on the chest with a harness and costs around $15. Be sure to angle the nailstripper so that the nails gravitate toward the exit slot. If you can't find one locally, write to South Coast Shingle Co., 220 E. South St., Long Beach, CA 90005; Roofmaster Products Co., POB 63167, Los Angeles, CA 90063, or McGuire-Nicholas Mfg. Co., 6223 Santa Monica Blvd., Los Angeles, CA 90038

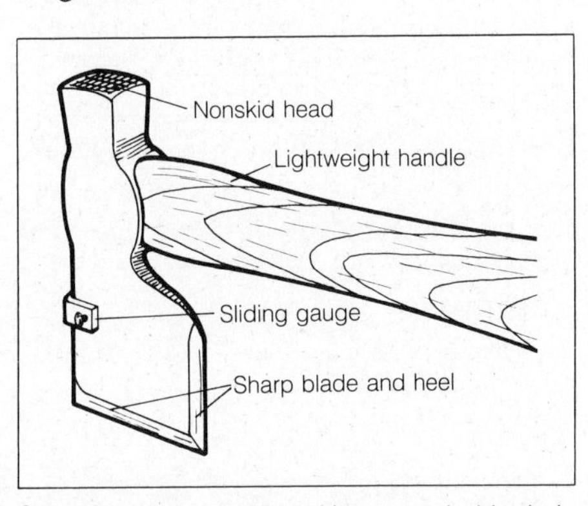

Shown is what to expect in a good shingler's hatchet. A high-quality tool will have most, if not all, of these desirable features.

Shingler's hatchet. Although you can get by with a hammer and a saw or chisel, this tool is highly recommended for wood roofing. Also called a lather's hatchet, it is often used by professional roofers for asphalt shingling. It's used for hammering nails as well as trimming wood shingles or shakes. If you buy one, make sure it has a sliding gauge for fast, accurate checking of exposure. A good hatchet has a cross-hatched, nonskid head at the blunt end, with a sharp blade and heel at the other.

Chicken ladder. This excellent device hooks over the ridge by means of broad 2x4s nailed to the top, to provide safe footing on steep pitches. You can make one yourself out of plywood and pieces of 2x2 or 2x4, as shown.

Ridge hooks. You can adapt a straight ladder for use on steep roofs using ridge hooks and 2x4s.

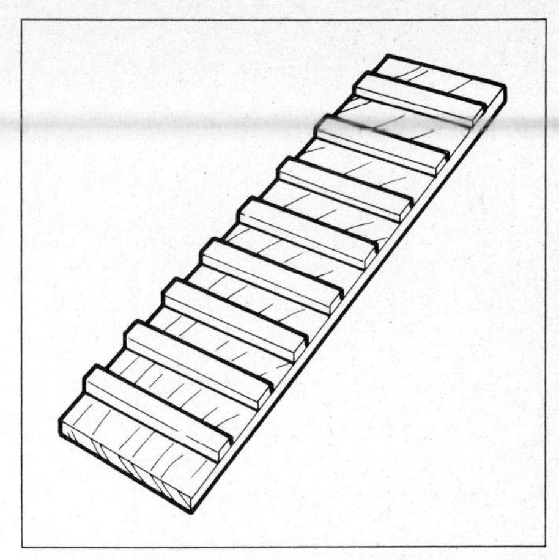

To make a chicken ladder, nail ¾-inch exterior plywood to 18-inch-long 2x2s (use 8d galvanized nails) spaced 12 inches apart.

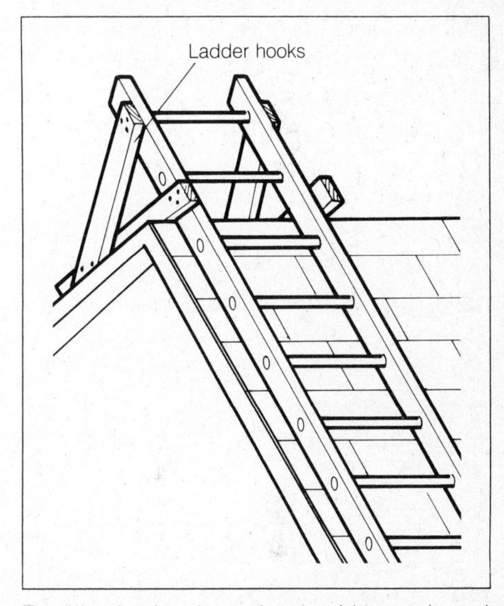

For ridge hooks, determine the ridge angle and attach 2x4 hooks to ends of wooden ladder using 12d common galvanized nails.

Safety harness. These can be bought at boat supply stores. The harness straps around your body and has a built-in hook that can be attached to any solid spot on the roof—or attach it to a nylon rope safety line.

Tin snips. When cutting metal flashing and edge strips, and also when trimming asphalt shingles, tin snips are very helpful. They are particularly necessary for odd shapes.

LADDERS—BUYING AND USING

Ladders were once little more than crude steps cut in upright logs, or simple constructions of wood timbers and cross pieces, notched and bound with thongs. Today, the range of ladder designs, types, sizes and materials range widely.

No matter what type of ladder you buy, study the various types and features available. Price should be a consideration, but not the main one. Safety and durability are the important features to look for. Don't be in a hurry, or let a fast-talking sales clerk steer you. Check everything yourself, looking for cracked and loose rungs, or other weaknesses before you leave with the particular ladder. Play it safe and purchase a heavy-duty ladder for this work, labeled "Type I", capable of supporting loads of 250 pounds. The ladder must be able to support not only your weight, but also the materials you will be carrying.

Never buy a ladder that is not clearly marked with the name of the manufacturer, the safe working length, or the seal of either the American National Standards Institute (ANSI) or Underwriters' Laboratory (UL). The absence of the seal does not necessarily mean poor quality, but the seal is an extra guarantee of safety.

Wood vs. Metal Ladders

If you will be working anywhere near electrical wires, choose a wooden model. Both aluminum and magnesium conduct electricity. When you have to use and maneuver a long, cumbersome ladder, however, the metal ladders may be safer because of their lighter weight and easier portability. As mentioned earlier, some people feel safer on a heavier, wooden ladder, especially in the wind—metal ladders may sway in windy conditions.

A good compromise, though expensive, is a metal ladder with fiberglass side-rails. These will not conduct electricity along the side-rails, which is the most dangerous aspect of use of all-metal ladders. It will have greater impact resistance than wood, aluminum or magnesium ladders.

Features of good wooden ladders. Quality ladders have one rung every foot. If the ladder is 16 feet long, there should be 16 rungs. The bottom rung should be about 7 inches from the ground. The top rung should be about 5 inches from the top end of the rails or legs of the ladder. Ladder rungs should be at least 1¼ inches in diameter. Stepladder steps should be about ¾ inch thick and fairly wide to support the soles of your shoes.

Ladder width is important. Buy a ladder that is at least 15 inches wide—even wider if possible. A quality ladder should be at least one foot wide at the top and fan out at the rate of about 1 inch per foot from the top of the ladder to the bottom of it.

Special ladder features include nonskid feet, interlocking rails, automatic extension locks, rope/pulley adjustment. Ladder "feet" should be equipped with safety boots or shoes, which can be ordered as an accessory.

Wooden extension ladders. Pulleys should not be less than 1¼ inches outside diameter, with minimum breaking strength of 500 pounds. Rungs must be

Side rail slide guide
Pulley
Rung
Rung lock
Side rails

Side-rail slide guides, which must attach securely, prevent the upper section from tipping as the ladder is raised or lowered.

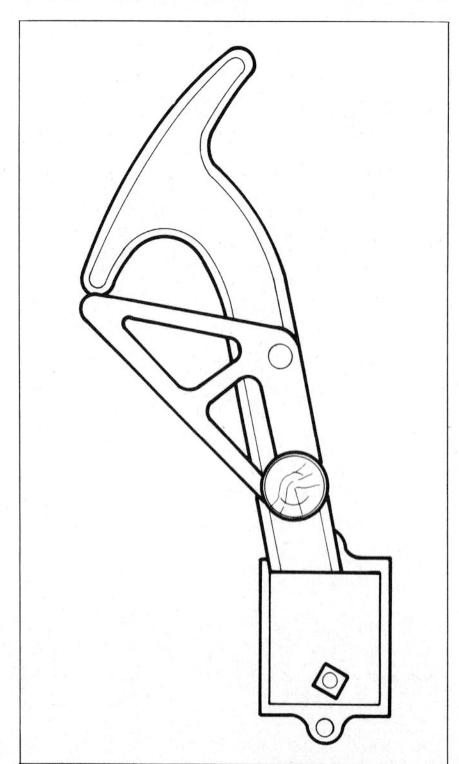

Rung locks on a ladder should be strong and rust-resistant, and have either a spring-lock system or a gravity-lock system.

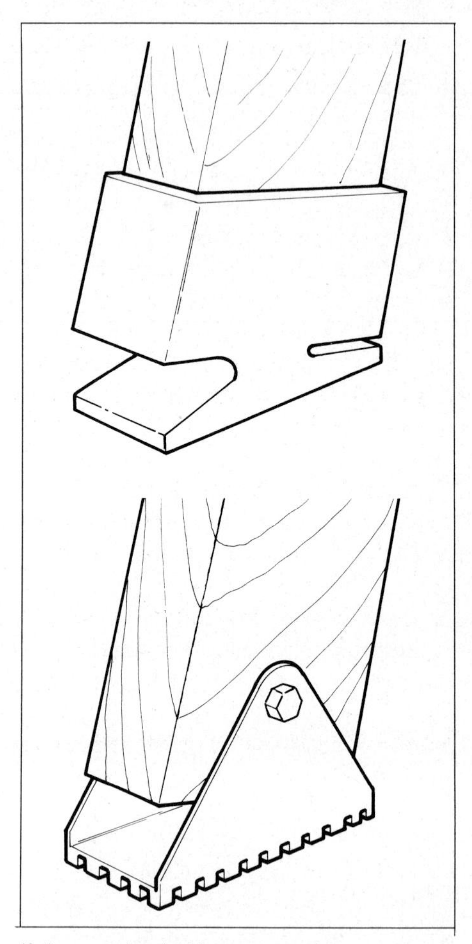

If the wooden ladder you buy does not have already attached safety feet, you can purchase and install one of the types shown.

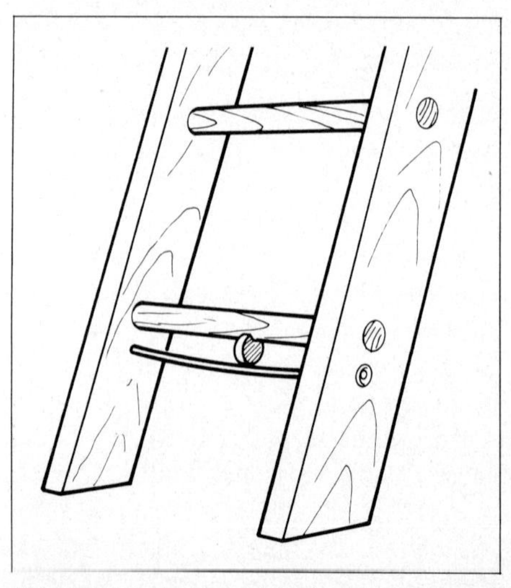

Although it is not essential, it is desirable that at least the bottom rung of a wooden ladder be reinforced with a truss rod.

round and made of hardwood, free from crossgrain, splits, cracks, chips or knots. They should not be less than 1⅛ inches in diameter, not spaced more than 12 inches apart.

Rung locks should be rust-resistant. Lock sections should be spring or gravity type. Slide-rail glides must be securely attached and placed to prevent the upper section from tipping or falling while the ladder is being raised or lowered. Although not essential, it is highly desirable for the bottom rung to be reinforced with a truss rod.

The minimum width for the upper section is 12 inches between rails; for the bottom section (up to 28 feet), it would be 14½ inches. To determine the proper ladder height for your house, see the accompanying chart. Since you will probably only buy one such ladder, make sure it is long enough to reach anywhere on the house. A 16-foot straight ladder, rather than an extension ladder, may suffice for a very low one-story roof. However, even an eave height of 9½ feet requires a 16-foot ladder. The top of the ladder should extend at least three feet above the eaves.

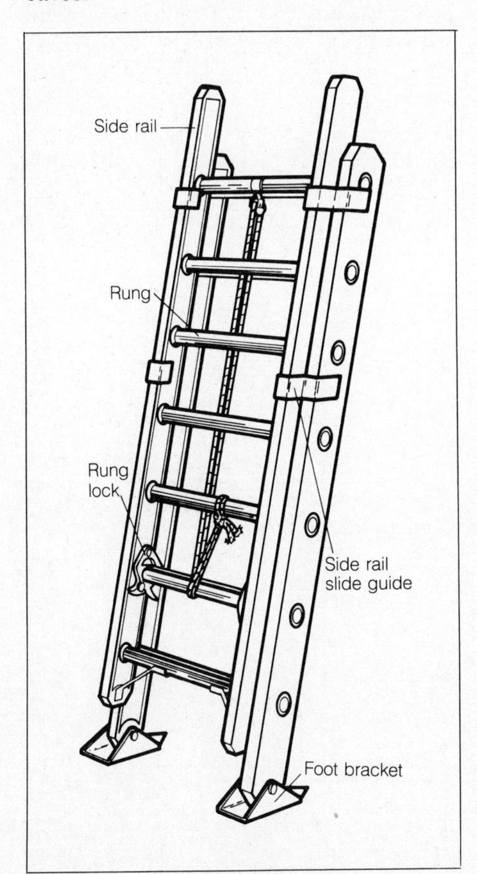

When carrying an extension ladder, lash it to keep the ladder from extending. Even when lashed, the ladder will be cumbersome.

Features of good metal ladders. Extension-ladder, pulley, rope and lock requirements are the same as for wooden ladders. Safety shoes should be attached by the manufacturer. Rungs can be completely round or rounded at the bottom with a flattened top. (A flat surface should be horizontal when the ladder is placed at the desired 75 degree angle.) Top and bottom rungs should be not more than 12 inches from the ends of the side rails, with a minimum of 12 inches between rails.

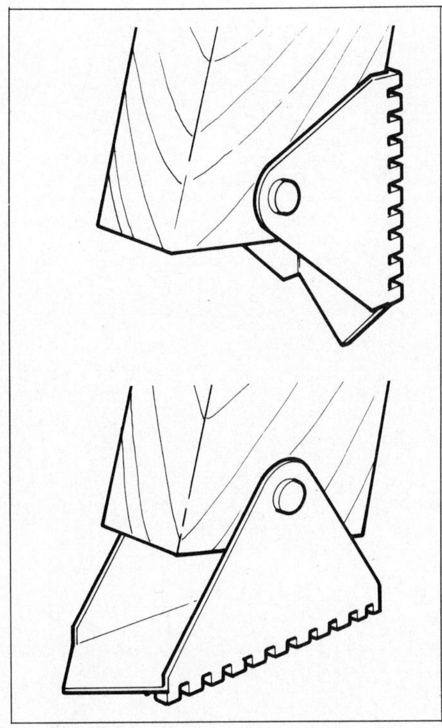

This is an alternative style of safety shoe or bracket. The pointed front edge can be forced into soft earth for greater traction.

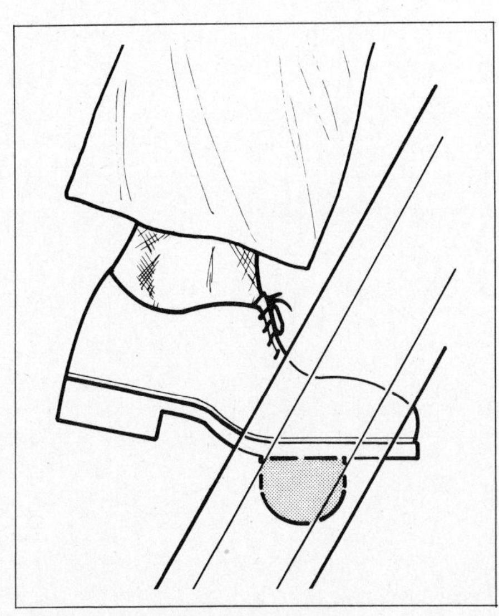

Rungs come completely round, or flattened. Flat tops are easier on the feet. The rungs must be level when the ladder is in use.

The width of the upper section should be at least 12 inches. Bottom sections vary, with a minimum of 12½ inches up to 16 feet, and 14 inches up to 28 feet. Rung braces are not required, but it is highly desirable that at least the bottom rung be reinforced with a metal rung brace. Buy ladders with rungs that will not twist and are peened in rails, if available.

Useful accessories. Safety-wall grips of rubber or plastic can be attached to the tops of extension side rails. These strips keep the top of the ladder from slipping on the surface against which it rests. Trays are handy for storing materials. If they are not attached, shoes (feet or brackets) should be purchased separately. Get at least three hangers for storage. Always store ladders, especially wooden ones, away from the elements, to prevent deterioration from rot, rust or insects.

Ladder Inspections

Inspect ladders prior to each use. Check for cracked or rotted rungs on wood

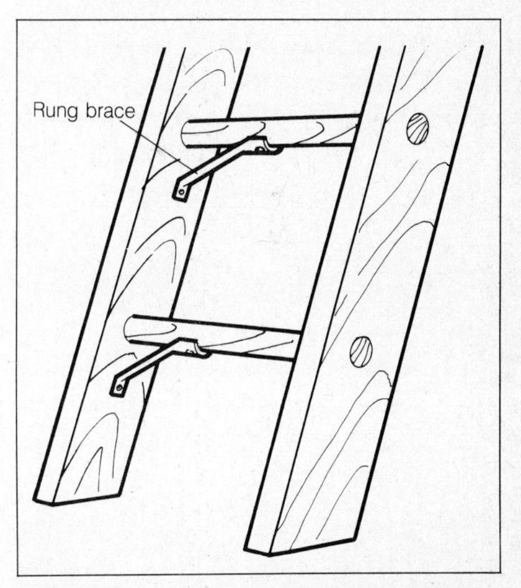

Braces are always preferred, and should at least be provided for the bottom rung, even if placement is not feasible elsewhere.

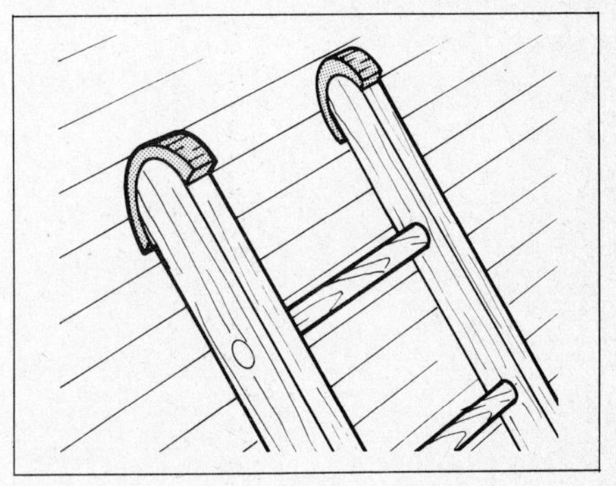

Use of safety-wall grips will help prevent a ladder from sliding once it has been placed against the side of the house.

ladders, and damage to metal ladders. Have any repairs done by the manufacturer or a competent repair shop, unless you are skilled in this type of work.

Oil any moving parts and tighten loose fasteners before each use. Check the rope and pulleys for wear and for breaking strength.

Do not paint wooden ladders. Paint can hide splits, breaks, and other defects. If you want to put a preservative on the wood, use a clear penetrating sealer. If you find a defect in a wooden ladder (or a metal ladder) such as a split, broken rung or a bent rail, buy a new ladder. Repairs seldom work.

Setting Up an Extension Ladder

(1) Move the bottom to its approximate final location.
(2) Brace the bottom against a surface on which it will not slide, like the side of the house.
(3) Grasp the uppermost rung, using both hands.
(4) Raise the top end and walk forward under the ladder, moving your hands along to push the rungs up.
(5) When the ladder is erect, adjust it to the desired location and lean the top forward to its resting point.
(6) Place the ladder at the correct angle to the wall, with the base at a distance of one-fourth the ladder's working height.
(7) Adjust the extension section of the ladder only when you can see that the rung locks are engaged and above eye level.

Double-check locks on ladder extensions. They hook automatically, but are not fail-safe. If a lock is damaged, do not use the ladder.

Setting up safely. If the angle is too great, the ladder is subject to strain that can break or bend it. If the angle is too small, the ladder is likely to slip backward. Always keep the ladder on a firm footing, with the feet level. Use a wide wooden board under the feet if the ground is soft. Place wooden shims under one foot if the ground is uneven. Whenever there is a doubt about the feet or top of the ladder slipping, lash it to a window or other solid support with strong rope. The 2x4 brace shown is one method. In an emergency, a second person can hold the ladder while you climb up and down. Never stand on the top rung of an extension ladder.

Use wide wooden boards (or the equivalent in smooth, even scrap lumber) to shim the ladder wherever the ground is uneven or soft.

Setting Up a Stepladder

Brace one leg of a stepladder against a wall whenever and wherever you can. When you open a stepladder, make sure it

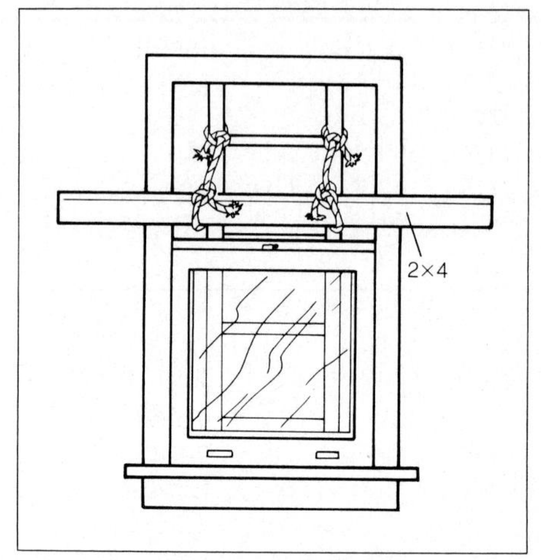

If there is concern about a ladder slipping, lash it to a window or other solid object. A 2x4 brace is one approved method.

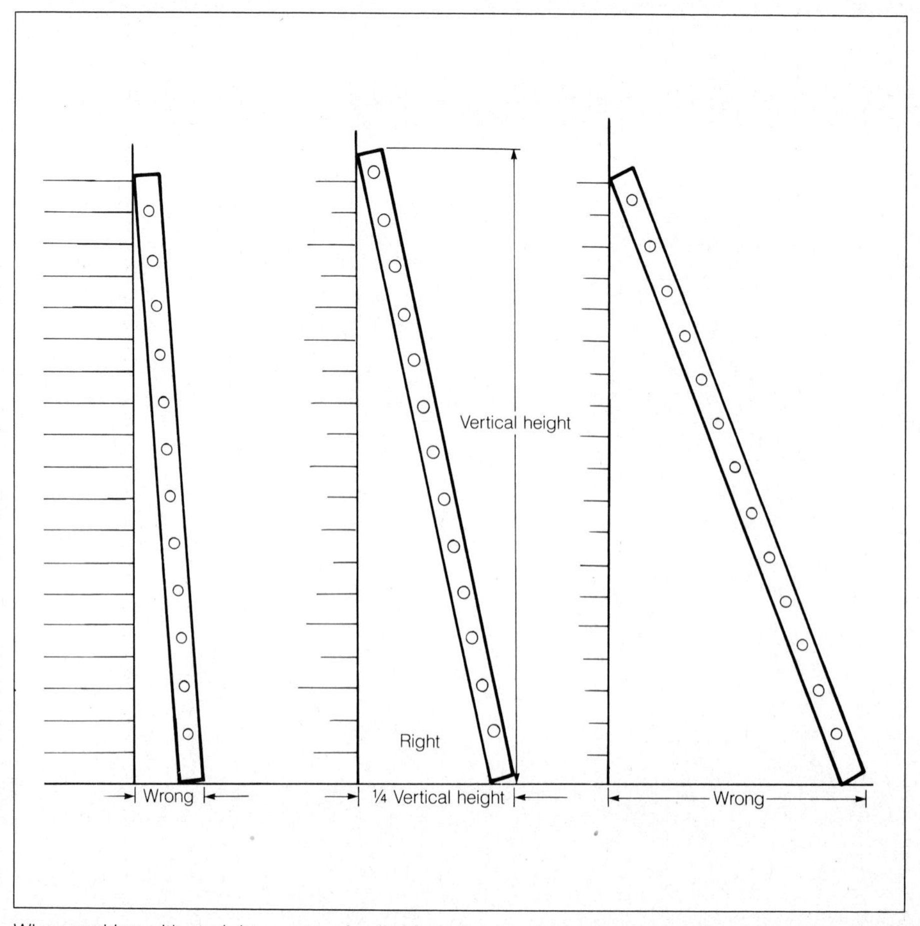

When working with straight or extension ladders, the correct angle requires that the feet be placed 1 foot from the house for every 4 feet up the side of the house (about 75°).

is fully opened, with the tray feature at the front of the ladder pulled completely down and locked. The metal straps running between the front and back parts of the stepladder must be fully extended and locked. Never stand on the next-to-the-top or top step of a stepladder.

Climbing and Carrying

When climbing any ladder, always face forward. As much as possible, carry tools in your pockets or in a tool pouch or belt. Keep both hands on the rungs. Take one step at a time, and step off one foot at a time while still holding the ladder. Never climb up so far that you have to bend down to hold the side rails.

Accidents often occur when the roofer tries to carry materials up to the roof. There are several ways to avoid this. The best is to have the supplier load the materials up there on delivery, using a scissors lift or other device. Sometimes, this is impractical due to the roof slope or lack of equipment. In that event, try to move the materials up in small packages that are light enough to handle. Even better, rent or build a scaffold or construct a wide ladder so that two people can carry bundles between them.

Begin making harness by wrapping a length of sturdy rope, such as a braided nylon climbing or nautical lifeline rope, around the waist—twice.

Each end of the rope is looped twice around the double thickness at the waist, then drawn back through the beginning of the loop in a modified Tarbuck knot.

The ends are looped in front, drawn to the back and crossed, then brought down over the shoulders and pulled through the doubled rope at the waist.

Now loop the end of the rope around both shoulder and waist lines, pulling the ends through the first loop to complete the knot—a modified Tarbuck.

To secure all the knots and prevent any possible slippage, tie a square knot and pull it against the Tarbuck knot. Leave enough slack for comfort.

Attach safety line—a very strong, sturdy rope that will withstand friction from the eaves, shingles and ridge—with a Tarbuck knot as shown here.

A roof is a dangerous place to work. Even a slight slope may cause a person to feel insecure. A rope harness will distribute pressure on the back should you fall. Leave little slack in the safety line; adjust slack each time you change position on the roof. Use sturdy rope. Climbing rope will withstand friction. It is expensive but a life-saving investment.

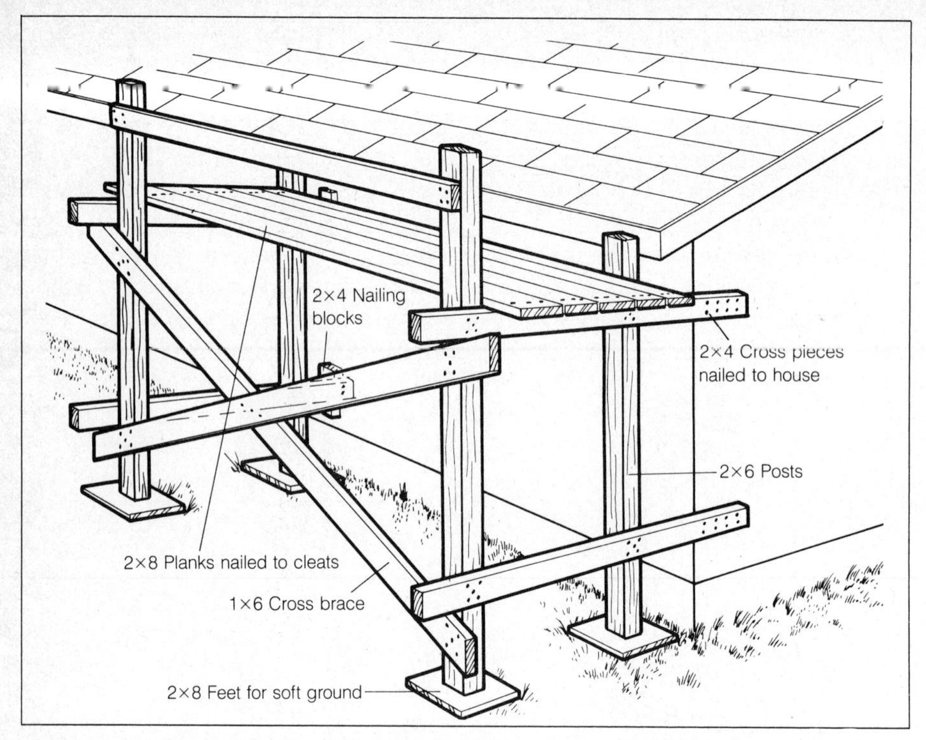

For safe working conditions, rent or build a scaffold. Details for building one are shown. For access, use a ladder or nail pieces of 2x4 or 1x6 to the end supports.

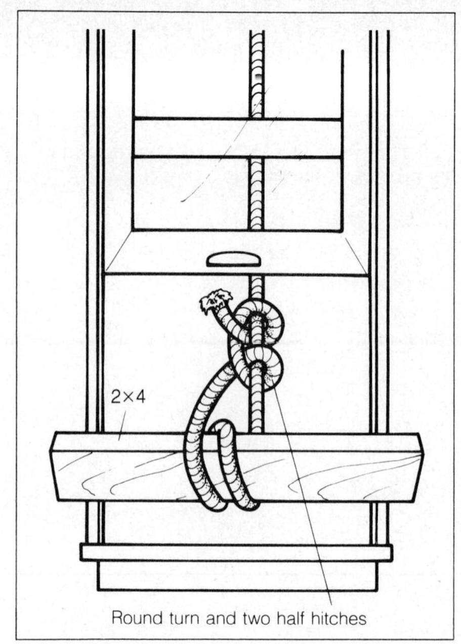

Secure a safety line to a board (2×4 or heavier) overlapping inside window frame. Stress on knots will tighten them.

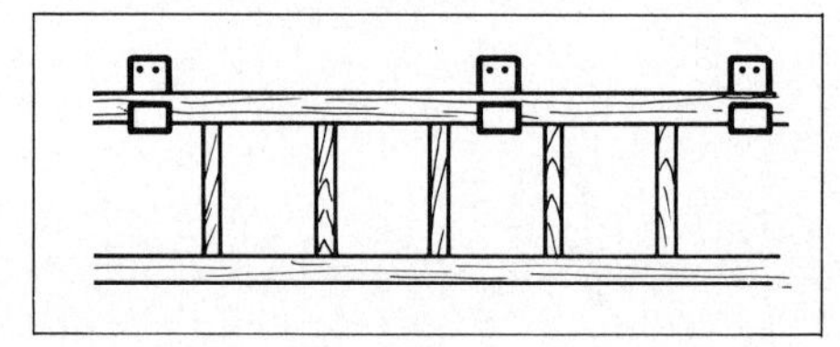

Store ladders, especially wooden ones, away from the elements. Hang them on the wall, using at least 3 special ladder hangers.

If you must carry materials up the ladder, and a rope hoist doesn't work or you do not have a helper to hand you the materials, lay the material across your forearms and use your hands to hold onto the rungs. Keep your hips inside the rails of the ladder.

Although it may mean a little more work because it involves moving the ladder, never stretch out too far along the sides of a ladder. Go back down and replace the ladder closer to the work. Do whatever possible to avoid electrical wires. If it is necessary to work close to live circuits, use only a *dry* wooden ladder. Wet wood can also be a conductor.

When working on a roof, make sure the ladder extends at least three rungs above the eaves of the roof. Do not set the ladder under the eaves and then attempt to climb over the eaves (and gutter) onto the roof.

Working Safely

On any roof, keep your eyes open and your senses alert. A slip of the mind is the cause of more falls than the slip of a foot. Be careful if an electrical storm is in the air. Disconnect the AC plug to your television set, and do not go near the antenna when it is wet.

A little healthy fear will keep you on your toes, but too much may paralyze you. Unless you are used to working on a steep roof, it is wise to attach a safety line, as described earlier.

Using cleats. When working on a slope with more than a four-inch rise, use 2x4 roofing cleats to provide firm footing. Make sure that the cleats are firmly attached to solid wood. The best way to do this is to drive 16d nails through the 2x4 into the rafters at three points. Plug the nailholes later with roofing cement.

6

SIMPLE ROOFING REPAIRS

If you have not already read Chapter 2 on roof inspections, do so before proceeding any further. Your inspection will tell you whether you need repairs—or whether your roof is too badly deteriorated to make repairs worthwhile. In the latter case, a new roof is needed, as explained in Chapters 7, 8, and 9.

It may be, however, that even if you decide that a reroofing job is called for, emergency repairs also are necessary. There is little point in making cosmetic or preventive repairs if a new roof is in the very near future. However, if there is already a leak that could cause interior damage, a repair should be made. It is also true that, even if you need a new roof, the money is not always available. When the roof need reroofing, but it looks like you may have to put it off until next year for financial or other reasons, carry out needed repairs in the meantime.

SHINGLE REPAIRS

Fixing Asphalt Shingles

The type of repair for asphalt and fiberglass shingles depends on what kind of damage exists.

Minor damage. If the shingle is cracked or has a few small holes, the simplest repair is to trowel some roofing cement over the damaged spots.

Wind damage. When a shingle has been lifted up and bent by the wind, a tree branch, or some other object, cover the area below it with a liberal layer of roofing cement. Hold it down with your foot for a minute; it should stay down. If not, drive in two roofing nails and cover the nailheads with the roofing cement.

Cupped or curled shingles. If many of the shingles are curled or cupped, it is probably time for a new roof. You can delay that evil day for a while by nailing the curled shingles back down again after dabbing some roofing cement underneath. Be prepared for a new roofing job in the near future, however.

To re-adhere loose but intact shingles, just place a quarter-sized dab of plastic roofing cement underneath each shingle tab.

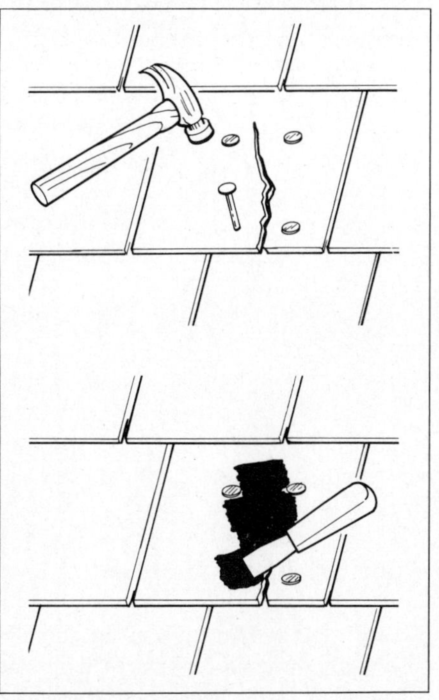

Split warped or cupped wood shingles and then renail on each side of the split. Cover split and nails with roofing cement.

Loose shingles. When shingles loosen or come off, it may be a sign of serious trouble. One or two missing nails can be replaced by new ones, troweling on some roofing cement over the nailheads afterwards. When a great many nails are loose, but the shingles themselves seem to be in good shape, there may be wood rot below, which causes the nails to loosen. If you can't get a good look at the wood from the attic, lift up the shingles far enough so that you can inspect the wood beneath.

Loosening or removing nails. To loosen the nails, run the straight end of a crowbar or nail puller under the shingles and lift them up. Be careful, however, if the shingles are brittle—this is more likely in cold weather. With brittle shingles it is easier and safer to remove the

To renail a loose shingle, lift up the shingle above. Nail so new fasteners are covered by the shingle above.

nails of the overlapping shingles by running a hacksaw blade under the shingles and cutting off the nails.

Replacing Missing Asphalt Shingles

As with loose nails, one or two missing shingles may not be a serious problem. If there are many loose nails, it must be considered a warning—an indication of more serious problems. You should be able to tell whether the shingles themselves have deteriorated. If not, suspect wood rot in the framing.

Visual compatibility. If the roof is highly visible from the ground, you must find a suitable, similar replacement. When the roofing is reasonably new, you should be able to find a shingle that looks the same. If possible, get one from the same manufacturer; however, you may not know who the manufacturer is or where to buy that product. Fortunately, most colors are quite similar even among different manufacturers. If you have a piece of shingle, bring it with you to the dealer. Remember that even new shingles will discolor some, and there are bound to be slight variations in color due to differences between batches. This should not be as big a problem as it would be with paint or tile, which are seen close up. Most people don't get near enough to spot the "sore thumb," although it may look that way while you are replacing it.

Remember that asphalt shingles come in strips, usually of three, although there could be two or four (or more). If only one "tab" is missing, you can cut out one tab from the replacement to fit, but it is better to replace the entire shingle.

Metal patches. If you intend to reroof in the near future anyway, you can use a piece of thin sheet metal flashing material to replace a missing shingle or part of one. The repair will stand out, of course, but this won't matter if the area isn't highly visible. Use two roofing nails to hold the patch in, and cover the entire area liberally with roofing cement.

Placement and nailing. If you are cutting out a tab or a piece of one, you may not have to remove any of the surrounding shingles. Simply cut out the new piece, smear the area below with roofing cement and lay the piece in place. Drive roofing nails at the corners and cover the junctions and nailheads with roofing cement.

As a temporary patch for a leak, insert a thin aluminum sheet underneath the 2 or 3 closest shingles, and apply roofing cement at the edges of the patch to seal the joints.

When replacing an entire shingle, remove all the nails from the damaged shingle. You will also have to remove or loosen the nails in the overlapping shingles above. Loosened nails may be better than removal if looks are important, because you don't have to drive in new nails through the shingles afterwards; just nail in the old nails again.

Slip the new shingle into line with the others and nail one inch in from each end and between tabs, 5/8 inch above the cutouts or 5 5/8 inches above the bottom (assuming five-inch exposure). If the shingles are not three-tab, align the two center nails 11 inches in from the edge nails.

Replace or renail the nails of the surrounding shingles. If any nails were removed or won't go back in place, you will probably have to drive new nails through the tops of the shingles. Cover any exposed nailheads with roofing cement.

Wood Shingles and Shakes

In general, the same comments about overall deterioration of asphalt shingles also apply to wood roofing. When wood shingles or shakes begin to curl up along the edges, succumb to rot, or have extensive splitting or loosening, you must recognize that repair only delays the day of reckoning. The best course of action is to install a completely new roof. In the interim, however, or when such problems are minimal, repairs can and should be made.

Repairing existing shingles. Loose shingles or shakes can be re-attached using roofing nails or the same nails as were used in the original construction. Drive two nails per shingle, about one inch from each edge. Warped or cupped shakes or shingles should be split with a wood chisel and renailed. Cover both the split and the nailheads with roofing cement. Cover small natural cracks with roofing cement, just as for manmade splits.

Sheet metal patch. When the split is more than about 1/4 inch wide, put a sheet metal patch under the shingle before renailing. (Tar paper can also be used, but will not last as long.) Cut out the patch with tin snips so that it extends about 2 inches out from all parts of the split. Nail through both the shingle and the metal at the same time, using as many nails as necessary to hold the patch—two to four, depending on the size and shape. Sheet-metal flashing material can also be used to replace a missing shingle or section, as for asphalt shingles.

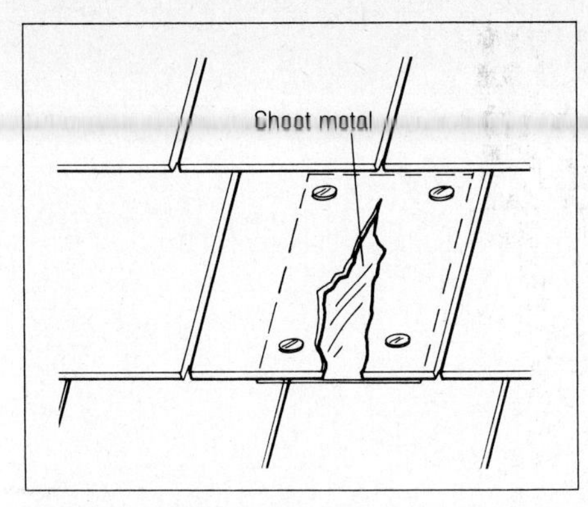

For splits that are more than ¼-inch wide, place a patch of sheet metal underneath the damaged wood shingle.

Inserting new shingles. It is better to replace a missing shingle with a new one, but the new wood may stand out as much as a metal replacement sheet because the older cedar will surely have changed during exposure to the elements. The steps for removal and replacement are quite simple.

(1) remove the damaged shingle by breaking it up with a chisel;
(2) remove nails with a hacksaw blade;
(3) once shingle and nails are out, measure the area to be patched;
(4) find a shingle to fit, or cut from a larger one;
(5) Tap the replacement into place under the shingle course above—use a scrap of wood as a buffer to prevent damage;
(6) nail at corners;
(7) cover nail heads with roofing cement.

Slate, Tile, Cement-Asbestos

Repair or replacement of a slate, clay tile, or cement-asbestos roof is a job for experts. Don't even trust asphalt or wood-roofing contractors for this job, since they will have had little experience with these other materials.

Fasteners. One job you may be able to do yourself is to replace the attachment devices for these roofs. Usually, these loosen or wear out long before the roofing itself. If you are very careful, you can re-attach the little hooks or whatever holds the roofing on. However, if the attachment is missing, you will have a difficult time finding a replacement. In this case you probably will have to hire a professional. In many areas, unfortunately, even that will be difficult. Competent repairmen are not easy to find.

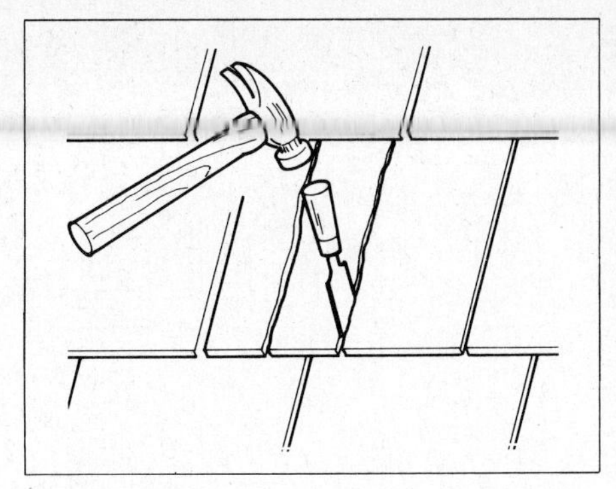

If the damaged shingle proves hard to remove, use a wood chisel to cut it into pieces before lifting it out.

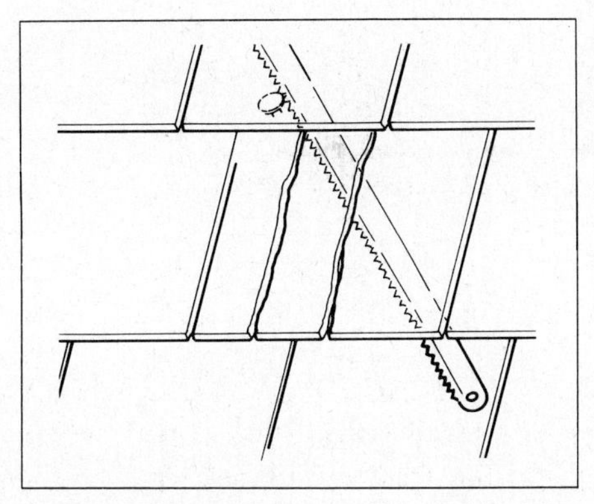

At the top of the shingle, cut off the nails holding the shingle in place by sawing through it with a hacksaw blade.

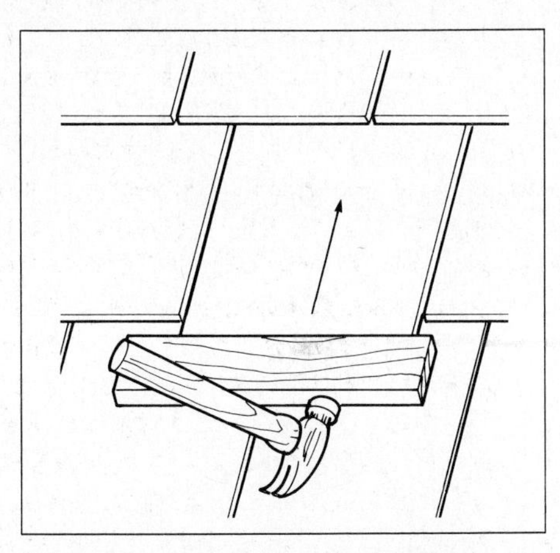

To insert the new shingle, drive it under the one above by hammering on the bottom. Use a piece of wood as a buffer block.

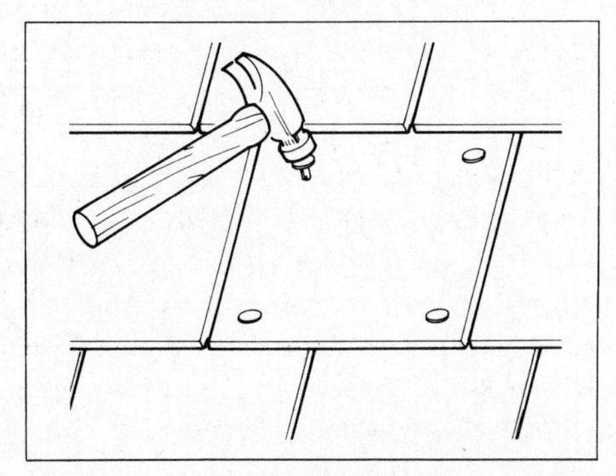

Nail the new wood shingle securely, as shown, in four places. Cover nailheads with roofing cement to ensure a watertight seal.

Spanish-style homes often have tile roofs. These roofs are very durable, but they are very hard for the homeowner to repair.

REPAIRING FLAT ROOFS

Built-up Roofing

Built-up roofing is used on flat and very low slopes. This consists of several layers of heavy roofing felt and hot asphalt.

Leaks. Most leaks in this type of roof occur around the edges or at a juncture such as a chimney. It may be difficult to tell whether the problem is with the roofing itself or with the flashing, but it doesn't matter because the cure is the same: apply a liberal coating of roofing cement in the problem area. Do not smooth it out. Extend the cement one to two inches beyond all sides of the problem area.

Fixing bubbles and splits. If the bubble or blister is intact and not split, and there is no apparent leak, it can be left alone. When the bubble splits or cracks, however, it should be fixed.

The methods for fixing blisters and damage on built-up roofing are similar. With a blister or bubble, the raised area should first be split down the center with a razor knife. Make a shallow cut, without damaging the layers beneath. Trim 1/16 inch to 1/8 inch off both sides of the split, because the paper will have stretched. If there are signs of moisture under the blister, open it further and let dry thoroughly before proceeding further. Then force roofing compound under the bubbled area with a trowel, putty knife or other flexible tool.

Drive roofing nails all along any cuts or loose edges, spacing them no more than an inch apart. Spread a generous layer of

This roof has continuous rafters spanning an opening for a light court. The line of the roof remains the same from ridge to eave.

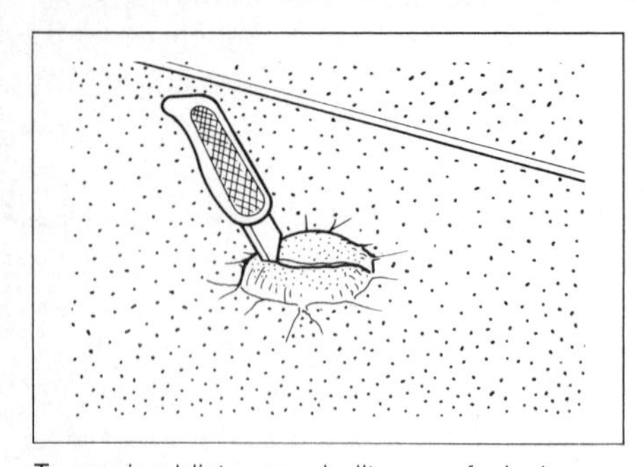

To repair a blister on a built-up roof, slash open the bulge with a utility knife and press the air out of the blister.

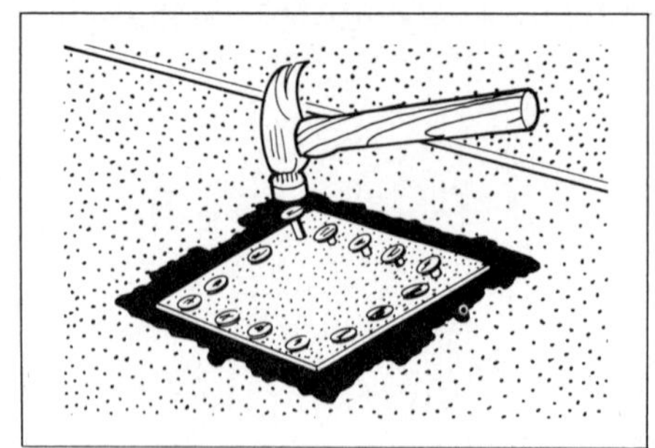

After coating the entire blistered area with roofing cement, cut patch of 90 lb. roofing paper. Nail edges at 1-inch intervals.

Spread another layer of roofing cement over the patch, extending the cement out an inch or two on all sides.

roofing cement along all cuts and edges, to cover all nailheads.

Now cut a patch from 90-lb. roofing paper large enough to cover the entire blistered or damaged area, extending about two inches beyond on all sides. Apply another layer of roofing cement to the underside of the patch. Nail it down along all edges, again spacing nails no more than an inch apart.

Spread another layer of roofing cement over the entire area, fanning it out another inch or so on all sides. Thoroughly cover all the nailheads and edges. Run yet another application of cement along the edges, just to be sure.

Gravel or stone covering. If your built-up roof has a gravel or stone covering over the roofing paper, sweep all of it away from the damaged or blistered area before working. Replace the stones when the roofing cement has completely dried.

Gouges or rips. When an area has been gouged or ripped, cement down any loose layers as above, applying roofing cement to the underside of any area put back down, and forcing it under any raised edges. Then proceed as above.

New gravel. Over the years, gravel on flat roofs tends to wash away with the

This home has a roof appropriate to the house style. It appears to be tile, but it is actually a one-piece material of tile-form curves.

rains. New gravel can be laid down but, before you replenish the gravel, put down a layer of aluminized coating. This fairly new material helps reflect the heat of the sun, in addition to acting as a sealant or cement. Some homeowners in the Sunbelt, in fact, apply this coating right on top of the gravel. It is cumbersome to apply that way, however, and the material will eventually wash away with the gravel.

Roll Roofing

Roll roofing does not ordinarily last much more than five years. The seams often open up and the nails begin popping out. If the roofing is on a shed or other out-building, you can coat the entire area with roofing cement or put another layer of roll roofing on top of the old one.

When the roll roofing is on your house, it should definitely be replaced with some other type of more durable and better-looking material. The new roof can be put on top of the roll roofing unless there are several layers of it already.

Replacing Large Sections of Shingles

It is unusual for a large area of shingling to deteriorate while the rest of the material stays in good shape, although sections can be blown away in regions with strong winds. Replacement of the entire area involves the same steps as for reroofing, which is covered in the next three chapters.

METAL FLASHING

Flashing is used wherever a roof meets another object, such as a wall, vent pipe or chimney. Flashing is made of thin sheets of copper, aluminum or galvanized steel. In most cases, repair or replacement is relatively simple. However, flashing around the chimney can be quite complicated and may require the services of a professional roofer. Instructions are given in Chapter 9.

Repairs

Most flashing leaks occur where the edge of the flashing meets the other object. Such leaks usually can be adequately sealed by applying a heavy layer of roofing compound along the edge. It should extend at least a half inch onto the chimney or other object, and a half inch onto the flashing; a one-inch or wider layer is necessary. Aluminum roofing or gutter tape can also be used to repair flashing. Gutter tape is for temporary repairs.

Crumbling mortar. Some leaks in chimney flashing are due to crumbling mortar where the flashing is embedded. Although you can replace the mortar if you wish, roofing cement will do the job if the deterioration is minor. Brush out all the loose mortar before applying the

Cut asphalt shingles to fit around the upper side of a vent stack. The cut must match the shape and conform to shingle overlap.

After shingles are in place, seal around the stack with roofing cement or other suitable flashing seal to prevent water leakage.

A tight seal on chimney flashing is crucial. After applying new roofing, carefully seal flashing against rainwater or snow.

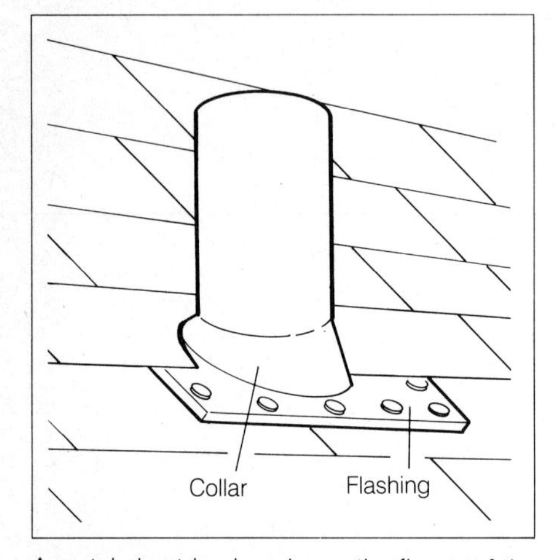

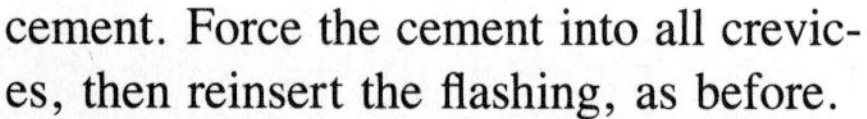

A metal sheet is placed over the flange of the stack vent to cover the joint between the stack and the roof. Fasten the sheet securely.

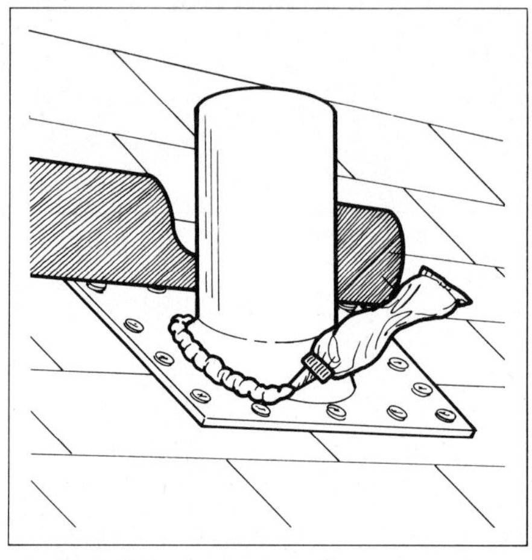

Use aluminized caulking compound in a caulking gun to seal joints and to prevent leaks between the collar and the flat sheet.

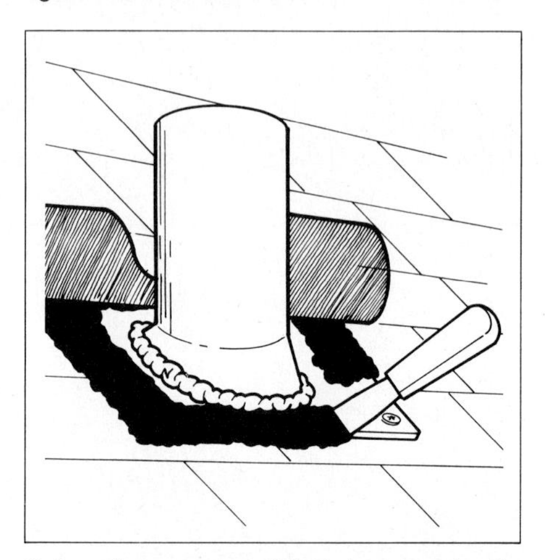

If there is a suspected leak between the flat sheet and the roofing, apply a generous bead of roofing cement to the joint.

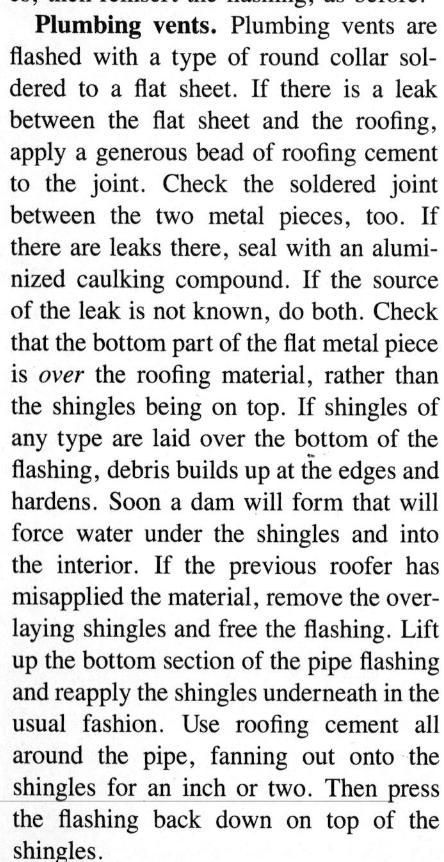

cement. Force the cement into all crevices, then reinsert the flashing, as before.

Plumbing vents. Plumbing vents are flashed with a type of round collar soldered to a flat sheet. If there is a leak between the flat sheet and the roofing, apply a generous bead of roofing cement to the joint. Check the soldered joint between the two metal pieces, too. If there are leaks there, seal with an aluminized caulking compound. If the source of the leak is not known, do both. Check that the bottom part of the flat metal piece is *over* the roofing material, rather than the shingles being on top. If shingles of any type are laid over the bottom of the flashing, debris builds up at the edges and hardens. Soon a dam will form that will force water under the shingles and into the interior. If the previous roofer has misapplied the material, remove the overlaying shingles and free the flashing. Lift up the bottom section of the pipe flashing and reapply the shingles underneath in the usual fashion. Use roofing cement all around the pipe, fanning out onto the shingles for an inch or two. Then press the flashing back down on top of the shingles.

Because shingles at valleys are subject to water runoff pressures, be sure that each shingle is firmly sealed at the valley edge.

Starter course shingles are cut to remove exposed edges. Cut the shingle with a matt knife (shown) for a clean, safe cut.

Replacement

Flashing replacing can be a complicated job, and in many cases it is a good idea to hire a professional. This is especially true of the flashing at the high side of the chimney, and sometimes at the side. Chimney flashing has two parts—base and cap or counter. Counter flashing is embedded in the mortar between the bricks. Some of this mortar must be removed and replaced when installing new flashing. In most cases, it is best to leave the old flashing in place and put new flashing over it. Remove any shingles which cover flashing as when replacing shingles. Then install new flashing as described in Chapter 9.

Sealing leaks. Leaks at the joint of the roofing and a wall or dormer may not be due to flashing, especially if it appears to be in good shape. Many builders assume that the flashing negates the need for caulking, but this is not true. Pull back the flashing and apply caulking compound along the joint below. Replace the flashing and seal with roofing cement.

This steeply pitched roof includes some essentially flat-roofed dormers to increase interior space and to accent the roofline.

Cedar shakes provide a dramatic and durable roof. Used on homes since colonial times, shakes and wood shingles have proven durability.

These shingles have been pressure-treated to withstand weather and resist fire. Wood shingles are attractive and have a good R-value.

Wood shingles must be overlapped to create a ridge. The exposed ends of the shingles should be away from prevailing winds and storms.

A steeply pitched roof would be difficult to repair. The owners installed a roof of terra-cotta-like shingles that are very durable.

Wood shingles are susceptible to moss when exposed to damp, such as these that are well shaded. Flashing was added to protect edges.

Although no longer commonly applied, metal roofing is available. This roof has been painted with a flat finish to prevent reflection.

The dormer in this roof would require flashing when reroofing it. The house was roofed in an older-style hexagonal shingle.

This house combines hips, dormers and gables. Eaves on main roof are curved slightly upward.

The main section of this home has a flat slope that meets the second gabled section. The far side has a curved (butterfly) slope.

A woven valley is an uninterrupted surface. Although not simple, it may be the best choice if a roof has many changes of plane.

7

PREPARING FOR REROOFING

If your present roof surfacing has undergone overall deterioration, you will need to start afresh. This generally means new roof surfacing, although you may need more than that.

DECIDING WHAT'S NEEDED

It is unlikely that you need a ''whole new roof'' in the technical sense. The framing and rafters rarely need replacing, although it is possible that dry rot, termites or some other condition will require removal of a structural member. If so, be sure to read Chapters 1 and 2 carefully so that you will know what it is you are replacing. Be especially careful if trying to replace any part of a truss roof. A professional carpenter may be your best choice if structural repair is necessary, particularly for trusses.

New Shingles Over Old

The first consideration when reroofing is to decide whether the old shingles must be removed. The main considerations are (1) the condition of the old material, and (2) whether the nailing surface is adequate for the new shingles.

Condition of old shingles. The new shingles must lay perfectly flat, meaning that cupped, warped or badly deteriorated older material must be removed. Roofs whose asphalt shingles have lost most of their mineral granules can offer a very good base for new shingles if the roof structure is sound. You can lay new shingles over old ones that are simply brittle or worn, but removal is necessary if they cause the new ones to heave or buckle. Missing shingles must be replaced to provide a level surface for the new shingles.

If there are only a few old shingles to be removed, they can be pulled up

A new roof requires a smooth, level surface. If old roofing is in poor shape or distorted, remove all of it before reroofing.

If you remove all shingles, check the sheathing for rot or other damage. Replace any sheathing section that is not sound.

If your old roof is basically sound, remove and replace any damaged shingles to provide a sound nailing surface.

When removing old roofing, remove and replace old flashing material at the same time to ensure a weathertight roof.

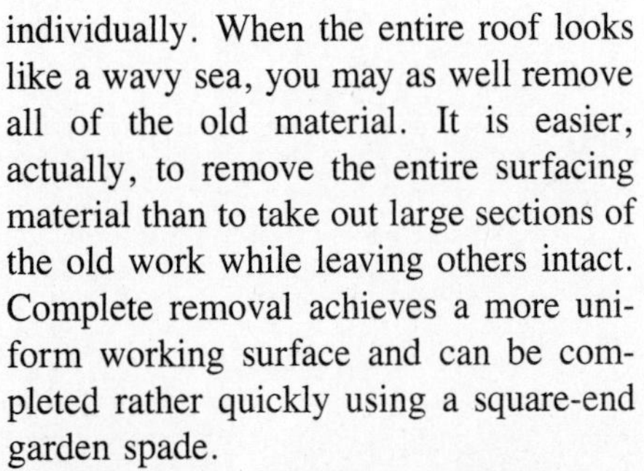

individually. When the entire roof looks like a wavy sea, you may as well remove all of the old material. It is easier, actually, to remove the entire surfacing material than to take out large sections of the old work while leaving others intact. Complete removal achieves a more uniform working surface and can be completed rather quickly using a square-end garden spade.

Wood plus asphalt shingles and shakes. You can lay wood shingles or shakes over asphalt shingles, if the asphalt shingles are in decent shape. You cannot lay asphalt shingles over wood shakes. The surface of wood shakes is too irregular for the asphalt shingles to lay flat. As discussed earlier, wood shingles and shakes are not the same thing. In many cases, you can install asphalt shingles over wood shingles, but this takes careful preparation, as outlined below. It may be simpler to remove the old shingles than to take these rather elaborate steps. On the other hand, the wood shingles, if left on, will provide a small degree of insulation (R-.87).

Nailing surface. The condition of the wood shingles is another important factor in deciding whether or not to remove them. They must provide good ''purchase'' (holding power) for the new roofing nails. If the shingles are soft and spongy, there will not be good anchorage for the roofing nails.

Weight problems. There is one additional factor to be considered, no matter what type of roof is being installed. The framing members and the decking must be strong enough to hold the weight of the new roof, plus the weight of the homeowner/worker while the roof is being installed. If there is only one layer of shingles currently on the roof, you usually can assume that one more application won't have much effect. If there has already been one reroofing done over the original shingles, the opinion of a contractor is advised.

KEEPING THE OLD SHINGLES

Preparing an Asphalt Roof

When putting a new roof over existing asphalt shingles, the old roofing must be smoothed out as much as possible. Nail down or cut away (with a utility knife) all loose, curled or lifted shingles. Loose or protruding nails should be removed or nailed back down—use new nails in new holes if necessary. Where the old shingles appear higher at certain points, there are probably lifted nails beneath. Locate these nails and nail them back down again through the shingle. If you have trouble locating the nails, drive in new ones to get the shingles back down again. Remove the old nails if they now protrude through the shingle.

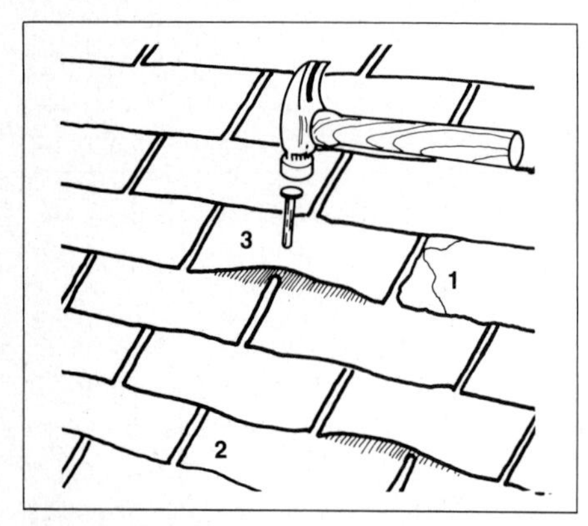

Nail down loose or torn (1) and cupped (2) edges, and insert new nails into cupped shingles (3) before applying new shingles.

Replacing wood or metal edging strips. When these are badly worn or warped, or ice dams are anticipated, the edging should be removed and replaced (see "Drip Edges and Flashing," below). Take off the old roofing to a point 24 inches beyond the line of the exterior wall. Sweep away all loose debris to provide a clean surface.

Self-sealing square-butt strip shingles over old square-butt strip shingles.

Shingles should be even at the rakes. Snap a chalkline along the rake and remove any shingles that overhang the line.

To maintain vertical and horizontal alignment, snap chalklines at regular intervals along the roof. Lines serve as guides.

The older asphalt shingles will cause the new shingles to lay unevenly. Although the new shingles will perform adequately, the surface will look uneven and lumpy, since the new shingles will conform to the configuration of the old ones. You may prefer to remove this type (generally single, "lock-down" or "staple-down" varieties) for cosmetic purposes. If you do keep these old shingles underneath, see Chapter 8 for reroofing procedures.

Preparing Wood Shingle Roof

If your inspection shows that wood shingles can be left in place, there are several things you must do to prepare the old surface for the new roof. Remove all loose and protruding nails, and hammer new nails in at different spots than the old nails. Renail loose shingles, and replace missing ones with similar shingle material. This does not have to be a perfect job, since the main purpose is to provide a uniform surface. If you cannot find an exact replacement, just do the best you can.

Badly warped or curled shingles should be split into sections, and each section

When reroofing with new shingles over old shingles, first replace the old drip edges with new strips at the rakes and the eaves.

nailed down separately. Any individual shingles that are badly deteriorated should be removed and replaced.

Rakes and eaves. If you live in an area where high winds are prevalent, it is a good idea to remove badly weathered shingles from the rakes and eaves. Cut off the shingles 5½ inches from each edge, and replace with nominal 1x6 lumber strips. Since the actual size of these boards is ¾ inch x 5½ inches, they

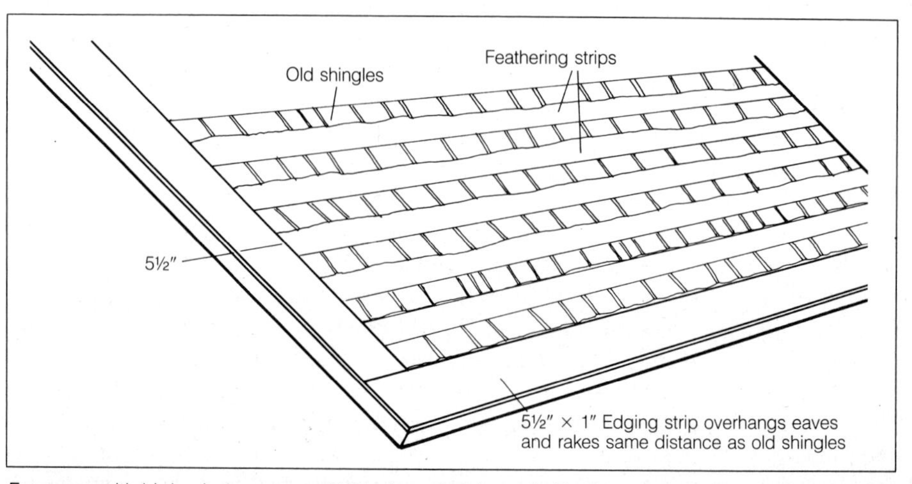

For areas with high winds, remove 5½ inches of old wood shingles around the perimeter; replace them with nominal 1x6 lumber strips.

To provide an even subsurface for new roofing, nail pieces of beveled siding against the butts of old wood shingles, with the thin edge of the siding at the bottom.

should fit into the space left by the cut-off shingles. The strip should overhang just as far as the shingles did. If necessary, cut back shingles to fit in the lumber. Add these strips to the sides also, as shown. Buy some 1x3 or 1x4 strips of beveled siding, and butt their wide edges against the bottoms of the shingle courses, with the thin edges against the top edges of the course below. This should feather out the surface so that the new roofing should lay relatively flat. Just before applying the new roof, sweep the surface thoroughly and remove all debris.

Preparing Roll Roofing for Asphalt Shingles

Slit any blisters or heaves in the existing roll roofing. Press the roofing flat to the deck and fasten down the pieces for a smooth surface. Take out loose or protruding nails; drive in new nails close by. If there are any lap joints that have completely separated, nail them down. Trim torn or damaged roofing so it has square or rectangular sides.

Inspect the deck for any knots or resinous pockets that should be covered with patches of sheet metal. Patch tears in the roofing with new pieces of roofing that are the same size as trimmed areas; nail them in place. Always sweep the deck clean before beginning application of new shingles. Proceed as for new construction.

Preparation of Old Built-up Roofing for Asphalt Shingles

For slopes between two and four inches per foot. Reroofing over built-up roofing should only be carried out if no insulation exists between the old deck and the felts. Take off any slag gravel that is in place, to leave a smooth, clean surface of underlying felts. (If this cannot be achieved, remove all the old roofing.) In either case, apply asphalt shingles on top of the felts just as for new construction on low slopes.

When there is rigid insulation under the built-up roof. Add a plywood nailing base to the old roofing. Follow the same procedures as for reroofing when the existing roofing has been removed. First, however, check the insulation to make sure that moisture will not be trapped underneath the plywood. Check local codes for the nailing base required.

Partial Removal of Asphalt Shingles

Often, especially in high-wind areas or where part of the roof is under a sap-dripping tree, only certain parts of the roof are in bad shape, with the rest of the roofing in relatively good condition. Eave and rake portions are those most subject to such localized damage. In such cases, you should remove the badly deteriorated roofing, but leave the decent parts intact.

Old shingle removal. The idea here is to provide a smooth, level surface for the new materials by building the bared areas to the same level as that of the other shingles. Beginning from the edges, remove the shingles from the worst regions with a garden spade or crowbar. Pound in or pull out any loose or protruding nails. Try to achieve a straight line at the juncture between the remaining shingles.

Re-covering the bare deck. Using new shingles or roll roofing, cover the bared area as neatly as possible, but you need not be meticulous about nail location, exact placement, or the existence of a weatherproof surface. The chief concern is to bring up the surface to the level of the existing shingles. All you really have to do is make sure this layer of shingles is uniform in height and won't shift. You should use the same weight shingles as on the rest of the old roof to ensure that they are at the same level.

If wood shingles are being removed, it is easier to replace them with an underlayer of 1x6 boards or 3/4 inch Exterior plywood than with new wood shingles. Do not use boards wider than 1x6. They might shrink or swell and buckle the finish roofing.

WHAT TO DO IF THE WHOLE ROOF IS IN BAD SHAPE

As discussed earlier, the decision to leave the old shingles or remove them is primarily based on whether there will be a smooth enough surface for the new materials. If the present roofing is in such bad shape that the new shingles will be bumpy and wavy, then a new surface must be provided.

New Decking Over Old Shingles

There is a third option between leaving on the old shingles and removing them entirely. Some people feel that it is easier to apply a new deck over the existing shingles. If there is already a solid deck, the new deck need not be as thick as the old. The idea is just to cover the bad roof and give a smooth new surface. To accomplish this, 1/4 inch Exterior plywood can be nailed over the old shingles.

Although 1/4 inch plywood does not add significantly to the roof weight, it is always wise to check with a contractor to determine whether the framing is strong enough to support the additional weight. It is usually not advisable to add new plywood if there is already more than one

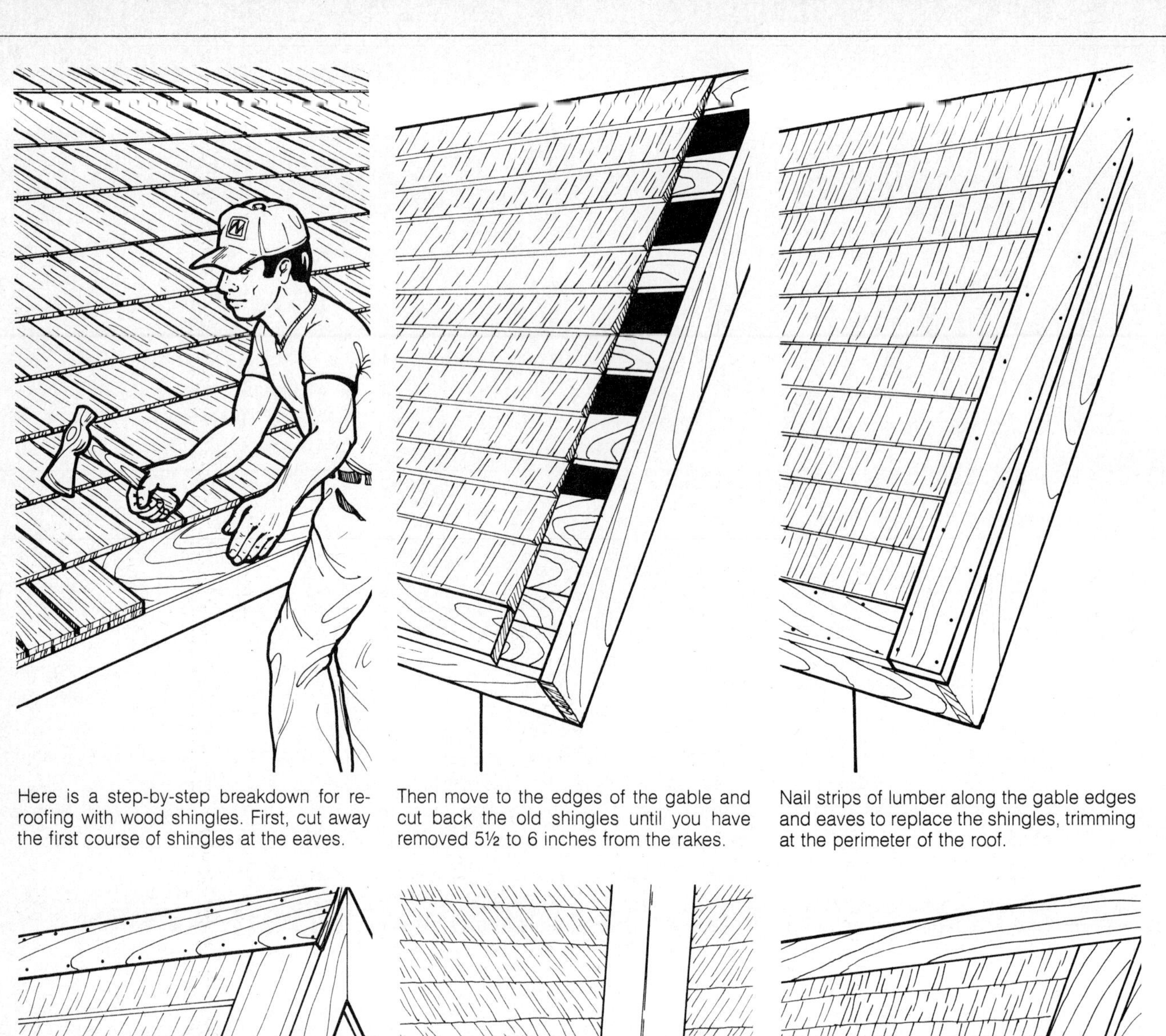

Here is a step-by-step breakdown for reroofing with wood shingles. First, cut away the first course of shingles at the eaves.

Then move to the edges of the gable and cut back the old shingles until you have removed 5½ to 6 inches from the rakes.

Nail strips of lumber along the gable edges and eaves to replace the shingles, trimming at the perimeter of the roof.

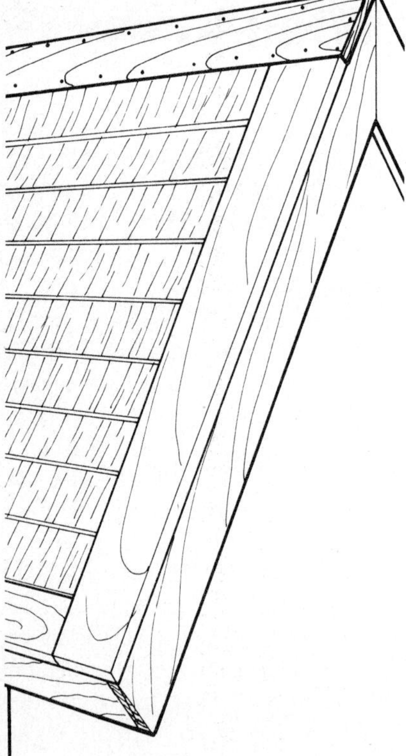

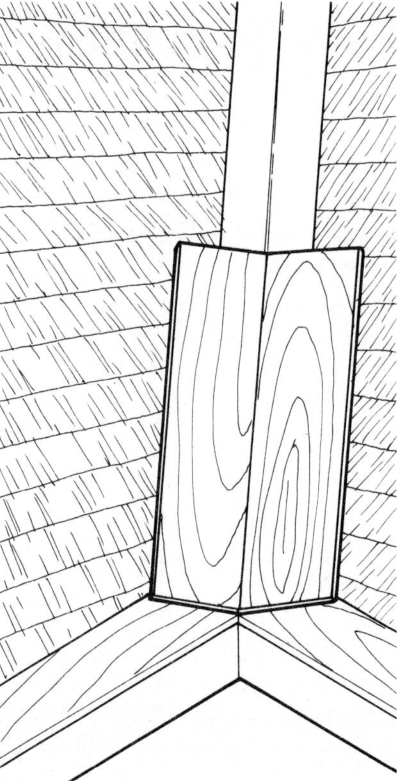

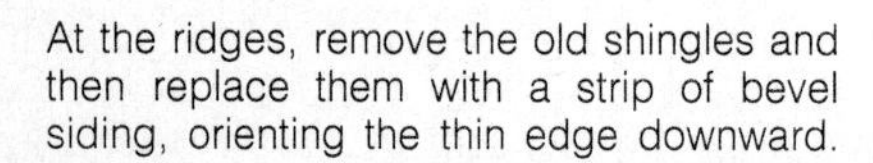

At the ridges, remove the old shingles and then replace them with a strip of bevel siding, orienting the thin edge downward.

Place additional strips of lumber inside each valley in order to divide the old metal from the new valley material.

Fasten the new shingles on top of the old shingles, using the old shingles as alignment guides. Use 5d rust-resistant nails.

The basic roof of this home was extended to provide a cover for a patio area. A roofline and eaves may be extended with relative ease if the original roof is flat or nearly flat.

These colonial-style buildings are suitably roofed with wood shingles that are appropriate in style, color, and texture to the wood siding.

Metal flashing material is laid over the felt for valley applications. The shingles will extend over the metal to prevent leaks.

Asphalt shingles overlap at the ridge and are then covered with a series of shingles laid on top of the ridge from end to end.

Wood shingles are most attractively finished with an overlapping wood-shingle ridge, but a metal-flashed ridge may be used.

The hips and ridge peak on this roof are finished with overlapping shingles for a neat appearance and a weathertight surface.

When the original roofline is not flat, the extension is added as a separate unit only partially supported by the house roof.

layer of shingles on the roof. When adding this type of deck, follow the same steps as those given below for applying new decking, but use 2 or 2¼ inch (6d or 7d) nails.

Removing Old Shingles

When your decision is to remove the old roofing entirely, the task ahead is not very difficult, but it will be messy and time-consuming.

The fast way. The most efficient method for removing asphalt shingles is to use a flat shovel. Remove all old nails before using the shovel. Slip the spade under the shingles at the eaves and push them up. You will also tear out the felt underlayment, which should be replaced before adding the new shingles. You may even dig up bits of the decking if you are not careful. If this happens, patch the decking with asphalt roofing cement. Work slowly near chimneys, pipes, vertical walls and other obstructions in order to avoid damaging the flashing. When removing soil stack and vent flashings that are no longer serviceable, use a pry bar. However, do not damage the old metal flashings, since they can be used as a pattern when cutting new flashings. Metal cap flashings at chimney (and other vertical masonry wall) intersections can be bent up out of the way if they have not deteriorated. Remove shingles in these areas carefully in order to prevent damage to reusable base flashings.

Repairing the deck. Inspect the deck to see if it is still sound. Repair the existing roof framing if necessary to strengthen it and to create a level and true deck. Replace any rotted, damaged sheathing. Cover large cracks, knot holes, or loose knots and resinous areas with patches of sheet metal; nail them to the sheathing. If there are loose or protruding nails, hammer them down or remove them.

The careful way. It is neater and cleaner to rip the shingles by hand, removing the nails with a crowbar or claw hammer.

Shingles and shakes. You may have to use the second method with wood shingles or shakes, whether you want to or not. A crowbar is the best tool for shakes and shingles, using the curved end to pull up the shingles and also to remove the nails. This is easier on the back than bending down with a hammer. You might try the spade method first, however, and see how well that works.

Open or slat sheathing. It is a distinct possibility that, when you remove old wood shingles, you will find that the deck is composed of nailing strips or "slat sheathing." This means that gaps of 1½ to 2 inches have been left between sheathing boards. One reason for this is to allow better ventilation and drying of the wood shingles from below. You should not have to wait until you remove the shingles to discover this situation; it can be checked from the attic.

If you are applying new wood shingles, you can leave the slats as they are, but if you are putting up asphalt shingles, you must fill these gaps. Strips of 1x2 or 1x3 furring can be used to fill the gaps, or larger nominal one-inch boards can be ripped to fit. It is not essential to get an airtight fit, but the boards should fit as snugly as possible.

Since fitting these boards can be a difficult and time-consuming job (the spaces are rarely uniform), it usually is easier to add a new plywood deck on top of the slats. Use ⅜ inch Exterior plywood to ensure a sufficiently thick nailing surface. Again, check with a contractor to see if the roof is strong enough to withstand the extra weight.

Safety. As with any roofing work, it is important to secure the area below the roof. Do not do any type of roofing work if little children are wandering below. Even with older children or adults around, a verbal warning may not be sufficient. It is wise to rope off the area and use warning signs. This is especially true while removing the old roof. It is a certainty that there will be all kinds of nasty objects cascading down from the roof when you remove the old roofing. In addition to roping off the area as out of bounds, cover any valuable bushes, flowers, shrubbery or other movable items such as statuary or lighting fixtures. Tarpaulins, preferably canvas, should do for shrubbery. Otherwise, improvise a more protective covering such as boards placed on garbage cans, temporary framing, or whatever your ingenious mind can come up with.

Disposing of debris. Remove portable items like outdoor furniture or portable lamps to a safe harbor. There will be a lot of old roofing material to be disposed of. Notify your sanitation department in advance so that they can cart it away. If they won't or can't, you will have to make a trip to the dump yourself. A rented truck or trailer may be necessary. If you have a leaf-shredding machine,

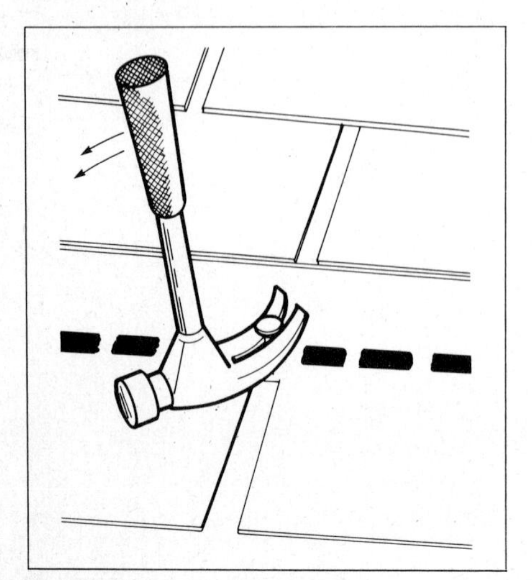

The neatest and best wood shingle removal calls for use of a hammer's claw end.

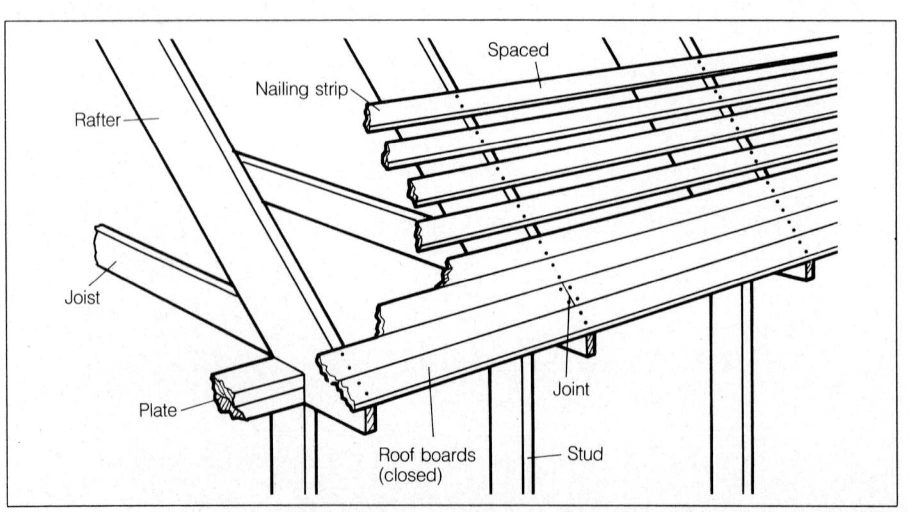

Roof boards under wood roofing may be spaced tightly together (at bottom), but many original wood roofs are nailed to spaced nailing strips, or "slat sheathing," (upper).

asphalt shingles and smaller, rotted wood scraps can be ground up and used for mulch. It may pay to rent one of these gadgets.

DECK REPLACEMENT

If the decking materials have rotted, they must be removed. Decks usually are constructed of 1x6 tongue-and-groove sheathing boards or of 3/8 inch Exterior plywood. The plywood should be sheathing grade or "B" (free of loose knots or other major defects) on at least one side.

Removing the Old Deck

To remove the old deck entirely, use a crowbar to pry the boards up where they are nailed at the rafters and at other framing. As discussed above, make sure that the area below is secure, and be very careful yourself. Since you will eventually be working on framing only, a slip or fall is much easier. You may want to consider hiring a contractor for at least this part of the job. Check that the attic area is cleared of anything valuable or breakable, since the materials will be falling down there as well. The attic and the outside perimeter must be off-limits to all members of the family until the job is done. Lock any doors leading to the attic and post a big sign at any scuttle hole.

Planning. Pay serious consideration to the weather. Check forecasts for a few days of clear skies, and be prepared with tarpaulins and/or heavy polyethylene sheets in case of an unexpected shower. Have all the necessary tools and materials on hand before beginning. Line up as many helpers as are available so you can get the job done as quickly as possible. Since plywood goes up faster, we recommend this over tongue-and-groove.

Installation of Decking

Plywood sheets. The plywood sheets can be installed as 4x8-foot panels, but they must be staggered so that the joints are at least one rafter apart. Start with half or third panels, alternating with full ones, as shown. The grain of the exposed plies should run horizontally across the roof. Use 8d hot-dipped galvanized nails every six inches at the ridge, valleys and rakes, and every 12 inches along the rafters. Leave 1/8 inch between panels to allow for expansion.

Tongue-and-groove. This type of sheathing is installed from the eaves up, cutting off the grooves of the first row of boards, and nailing with the tongues up. Use the same nails as for plywood. Drive in two nails where each board meets a rafter, and stagger the boards so that no two adjacent boards are joined at the same rafter, as shown. Avoid short boards at roof ends, especially where there is a substantial rake overhang. Make sure that end boards hit at least two rafters, and three or more if there is an overhang.

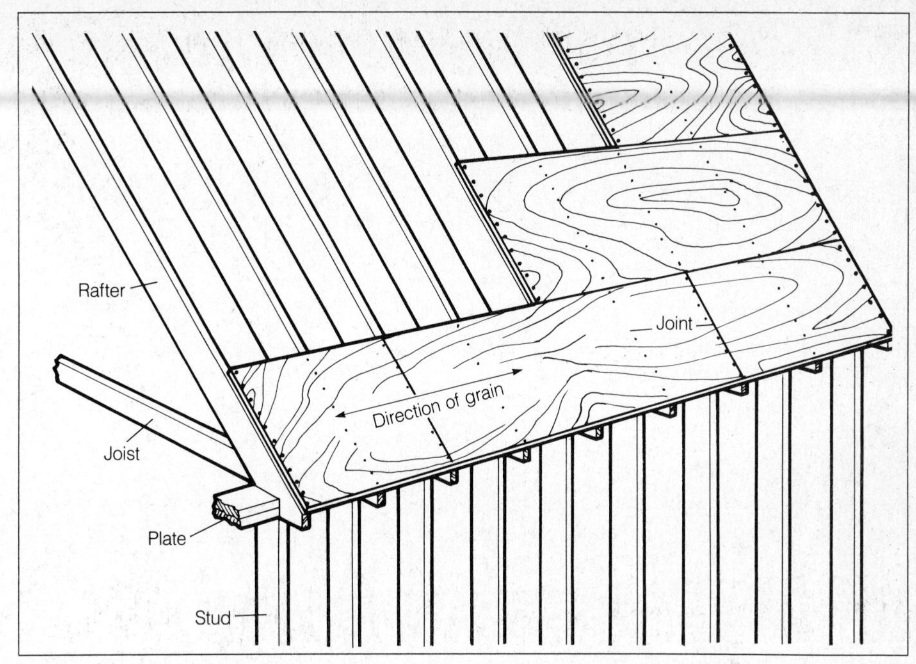

Plywood decking is the easiest to apply. Make sure that the outside grain runs horizontally, and that the joint lines are staggered as shown.

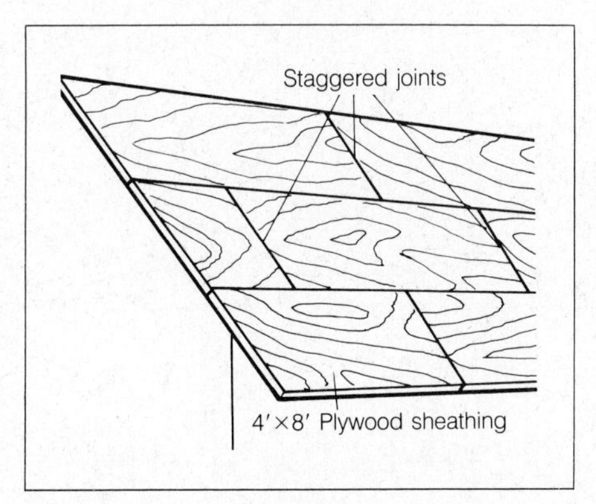

The staggered joints strengthen the deck and prevent leaks from following a continuous seam, which causes moisture damage.

The most efficient way to apply roof sheathing boards is to line them up ahead of time. After nailing one board in, slide the next one down, tapping with a hammer, to seat the tongue into the groove. (Use a block of wood so that the hammer does not damage the tongue.) You do not need to cut off the ends of the boards until the job is done. After all the boards are in place, use a circular saw to cut off the board ends to the desired overhang (or flush, if indicated).

Hips and valleys. Where hips and valleys are involved, save time and difficult on-roof cutting by determining the exact cutting angle before starting and then cutting to fit ahead of time. Take a saw with you, however, because some adjustments may be necessary. The fit should be especially tight in valleys. Use caulking at these joints.

Around the chimney. When sheathing around a chimney, leave 3/4 inch clearance on all sides to avoid a fire hazard. Rafters and headers should be at least two inches away from chimney masonry. The open areas will be covered with metal flashing.

Partial Deck Replacement or Repair

Your roof is more likely to need only a partial sheathing replacement than an entire new deck. If the wood is rotten, saw out the bad section to mid-point of the rafters on each side. A circular saw, set to the proper depth, can be used for large sections. Otherwise, a saber or keyhole saw can be used to cut off the wood alongside the rafters, and a chisel used to take care of the remaining wood. Be very careful not to cut through rafters or other framing.

When the present sheathing is tongue-and-groove, it is more difficult to cut out a single board, or the bottom and top pieces of a section. Remove the tongue of the last board that will be replaced using a power saw to cut along the top of the board. Then try to pry up the board. If it will not lift, then saw along the bottom of the board to remove the tongue of the good board below.

If replacing a single board when the tongue of the good board below is intact, cut off the tongue of the replacement board, and slip the new board into the groove below, dropping the top of the board into the space.

Using Metal Patches

When the old deck is stripped of shingles and underlayment, or when a new deck has been installed, inspect it carefully for defects which may allow penetration, such as holes or loose knots. Cover these up with patches made of approximately 26-gauge galvanized steel, copper or zinc.

The resin which sometimes seeps out of the wood at cracks or knots will react chemically with roofing materials and damage them. Therefore, all knots or defects in the wood which have a "wet" or resinous discharge should also be covered with a similar metal patch.

If the top surface of the deck is damaged or rough, without any sign of rot, the wood does not have to be replaced. Very rough surfaces should be planed or covered by a piece of metal, as for resinous or loose knots.

Underlayment

Asphalt-saturated felt (one layer of 36-inch-wide 15-pound material) is laid horizontally across the roof, parallel to the eaves. It should lap two inches over the layer below, and four inches over the end of the preceding strip. Most roofing felt comes with preprinted white lines for proper overlaps. When reroofing over old asphalt shingles, a new underlayment is not required. Never use tar-saturated materials, tar-coated felts or laminated waterproof paper. These might function as vapor barriers, trapping moisture between the roof deck and the underlayment.

Over old wood shingles. Local codes sometimes specify No. 30 asphalt-saturated felt for use over wood shingles.

On the deck. Professional roofers apply the underlayment as they shingle, working just a little ahead of the shingles.

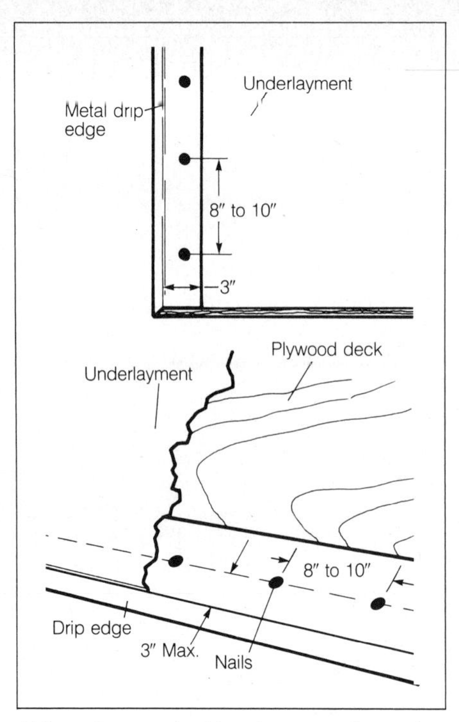

At the rakes, apply drip edges over the underlayment, extending 3 inches onto the felt. Nail drip edges every 8 to 10 inches.

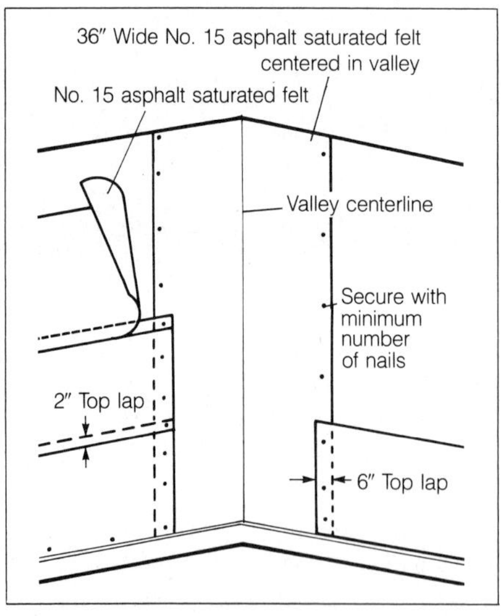

Begin roofing job by laying No. 15 asphalt-saturated-felt down the valleys of the roof and nailing just enough to hold it in place.

You may find this a little awkward and will probably prefer to put down all the underlayment at once. Nailing is not necessary, but the felt will slide and blow around unless you hold it down somehow. Most people use a few nails here and there in strategic spots, but the material tends to rip away from the nails due to wind. A few bricks, stones or other weights may do a better job until the shingles are laid over the felt.

Use only enough nails required to keep the underlayment in place. Locate end laps in succeeding courses at least 6 feet from any end laps in preceding courses. At hips and ridges, place the felt so it covers the hips and ridges from both sides and laps by 6 inches. If the roof meets a vertical surface, extend the felt up the surface by 3 or 4 inches.

Valleys. Underlayment should be laid in line with any valleys. If you are roofing over a bare deck that contains valleys, lay a strip down the valley before applying the horizontal layer. Center the felt over the joint. Insert just enough nails to hold it in place. Then apply the horizontal felt layer, overlapping the valley felt by at least six inches. Nail or weight the felt down at the ends and cover with valley flashing as soon as possible, before the underlayment blows or rips away.

Drawing in Chalklines

When reroofing over old shingles, the existing material serves as a guide for the new shingles. For new construction, however, chalklines provide the guidelines that ensure correct application of the shingles.

Horizontal chalkline, parallel to eaves. Measure and mark off the necessary distance on the roof; do this in the middle and at each end. Place one nail on the marks at each end. Stretch the chalkline between the nails and use it to check the alignment of the mark in the middle. Resnap the line from the middle nail to check the positions again.

Vertical chalklines. These not only are important for alignment of cutouts from the eaves to the ridge, but also for alignment of the shingles at the sides of a dormer. This keeps shingles and cutouts even, so they meet above the corner without gaps or overlaps. For long runs, snap the vertical chalkline so it falls in the center of the run. Apply shingles to the right of the line and then to the left. Also check the horizontal chalklines as the shingles approach the ridge; keep the upper courses parallel to the ridge.

Replacing Drip Edges

Drip edges are metal or vinyl strips that are used at eaves and rakes to prevent water from draining back under the wood. This prevents dampness that causes rapid rotting and wood decay. Drip edges are a small but very important part of the roofing process. If neglected, you can be sure that the roof edges will deteriorate much faster than expected.

If the old drip edges are in good shape, they can be left on at the eaves. Existing drip edges should be removed at the rakes when old shingles are removed, however, because new ones are installed on top of the new underlayment.

Before applying the underlayment at the eaves, nail the drip edge strips so that they cover the front of the deck edges, and are bent back three inches onto the top of the deck. Drive in 4d hot-dipped galvanized nails every 8 to 10 inches on top. Rake-edge strips are applied in the same way, except that they are nailed on top of the felt underlayment.

Repairing Flashings

Flashings are discussed at length in Chapter 9. One word is relevant here, however, concerning existing metal flashings, such as those around a chimney. Metal flashings do not ordinarily deteriorate; they can be replaced and used again.

Some flashings are shingled over, some are not. Shingle over those flashings that are presently covered with shingles, and put shingles under flashings where they are currently underneath. In the latter case, the flashing can sometimes be raised or bent up and used again. If not, a new flashing will be in order.

In any case, check all flashings carefully to determine if they need replacement. In most cases, defective flashings are simply bent or loose. A common cause of faulty flashings at the chimney is loose mortar where the metal is fitted between bricks. "Tuckpointing" (see Chapter 9) will tighten them up again. Vent pipe flashing usually is loosened so that shingles can be laid under the bottom section.

Eaves flashing. In order to prevent ice dams, which force water underneath the roofing, causing damage to interior walls, ceilings and insulation, you must add eaves flashing wherever there is a chance of ice buildup. The choice of flashing material and its width will depend upon the conditions that occur.

Slopes of 4 inches per foot or more. Place smooth, coated roll roofing (not less than 50 pounds) so it is parallel to the eaves and overhangs the drip edge by 1/4 to 3/8 inches. Install the strip so that it reaches 12 inches beyond the line of the interior wall. If you need another flashing strip to reach that far, place the lap in front of the line of the exterior wall. The overlap should be at least 2 inches. Cement the entire length of the horizontal joint of the overlap. At the ends, laps should be 12 inches and, again, cemented on meeting faces.

Low slopes. Use two layers of No. 15 nonperforated asphalt saturated felt to cover the deck. Nail a 19-inch-wide underlayment strip along the eaves. It should overhang the drip edge by 1/4 to 3/8 inches. Add a full 36-inch-wide sheet on top of the starter strip, with its long edge placed along the eaves so it completely overlaps the starter strip.

For the rest of the courses use 36-inch-wide sheets positioned so they overlap each preceding course by 19 inches. To fasten each course use only as many nails as are necessary to hold it in place until you apply the shingles. Create end laps that are 12 inches wide, placed 6 feet or more from end laps of the preceding course.

When constructing the eaves flashing, cover the surface of the starter strip with a layer of asphalt plastic cement. Embed the first full course into the cement.

With the first course in position, apply

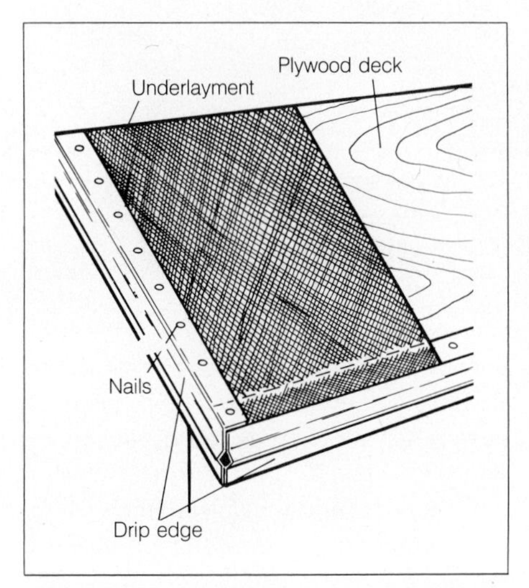

Metal or vinyl drip edges should be applied to the eaves before putting down the underlayment over the bare deck.

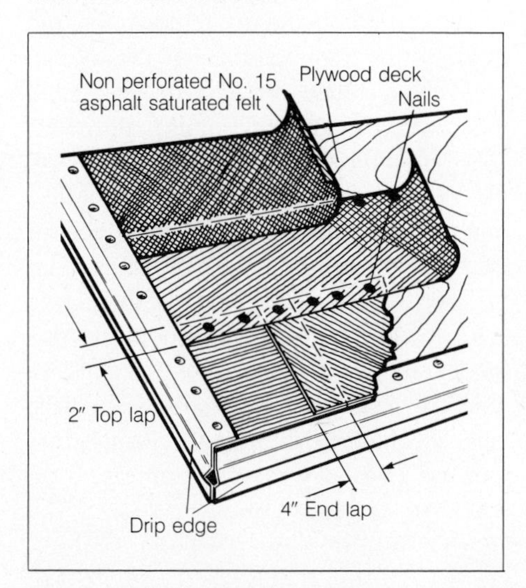

Place the first strip of underlayment at the eaves and overlap with the next strip by 2 inches. Seams overlap by 4 inches.

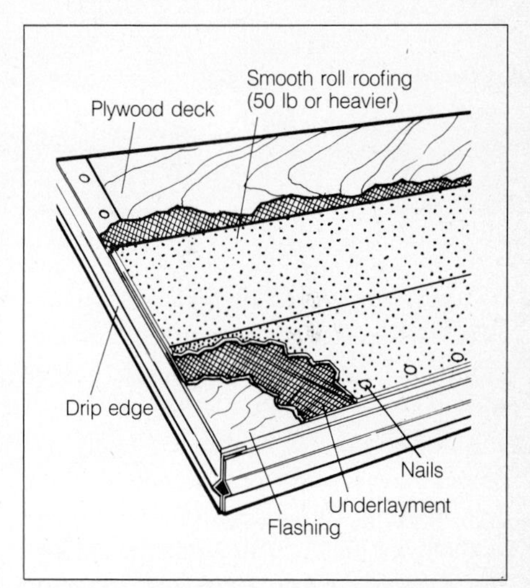

Eave flashing overhangs the drip edge and runs back on the roof to at least 12 inches inside the interior wall line.

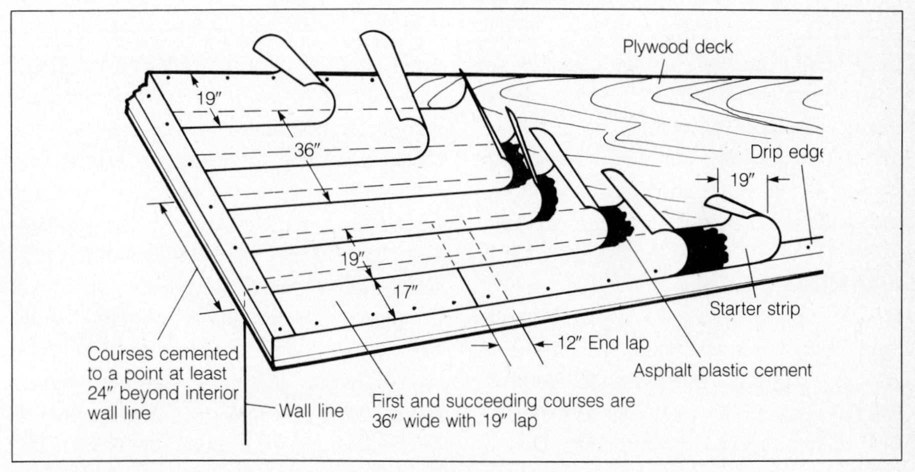

Underlayment for strip shingles on a low slope must overlap at least 19 inches to inhibit water backup under the shingles. A low slope drains slowly.

This tin roof has lasted over fifty years. Laid over wood sheathing and roofing felt, the roof reflects heat and offers insulation value.

a coat of cement to its upper 19 inches. Press the second course into the cement. Continue the procedure for all the rest of the courses that are within the eaves flashing area. Be careful to apply the cement evenly to keep the overlapping felt floating completely on the cement without its touching the underlayment in the course beneath. Try not to apply excessive amounts of cement, because it can cause blistering.

Severe icing. The strip of roll roofing should extend at least 24 inches further than the interior wall. Cement the flashing strip to the underlayment. Apply asphalt plastic cement uniformly, at a rate of about two gallons per 100 square feet. Embed the flashing strip by firmly pressing it into the cement.

FRAMING REPAIR

If any part of the roof framing shows evidence of deterioration, we definitely advise seeking professional counsel. Rarely does a "roof fall on your head," but it can and does happen. This is particularly true if there are signs of rot, decay or insect infestation.

Localized Damage

If there is localized damage, a cracked rafter or some other condition that seems to be confined only to one or two parts of the framing, the damaged part can be repaired. Because of the delicate engineering concepts in roof construction, do not try to remove the damaged part—at least not until you have provided a replacement. It is possible to damage the entire structure by removing one section of it.

Rafters. When a rafter needs replacement, study how it was cut and installed, and make a new one of similar material. Rafter angles must be precise, so first create a tracing or template. Use a protractor or a device sold in hardware stores to duplicate angles, and cut the replacement to the exact same angle. Put the new rafter right next to the old one, nailing it to the existing one with 12d nails. Use 8d nails to toe-nail the rafter to the ridge, joists, eave boards and wherever else the rafter meets other parts of the framing. If the deck is exposed, locate the new rafter. Nail through the deck to the new rafter. Do not nail through finish roofing into the new rafter. Do not cut away the old rafter. It won't do any harm, and helps keep the new rafter in place because you nailed into it.

Other framing. Unless there is a compelling reason for removing it, leave the old framing where it is. It won't hurt anything, and it is usually an aid to keeping the new framing in line.

If you do intend to remove the old framing for some reason, you do not nail into it, of course. Be sure to get a solid bond with the adjacent framing, however. The only time you might want to cut away the old framing is when you are modifying the attic to create new living space, or adding a skylight, dormer or similar project. This is not the topic here (see Chapter 11), but we must emphasize that no matter what the project, be positive that the new framing is solidly in place and adequate to support the load before you cut away any existing framing.

8

BASIC ROOFING PROCEDURES

The difficulty and cost of reroofing depends upon the type of roof involved. A simple gable roof on a one-story home may have only two planes, with no need for valley work or complicated flashings. If you are laying new strip shingles over existing strip shingles, you can begin work immediately with little or no preparatory work and no underlayment.

More complicated roofs require more complex tasks, and more instructions. Roofs that are badly deteriorated need more preparatory work. Those with L-shapes, dormers, additions, and extensions, also have more valleys, flashings, and similar additional work.

In the majority of cases, however, reroofing involves laying new asphalt shingles over old ones. This is not difficult in execution, but it is time-consuming. It requires patient, careful work. The basic procedures are given in this chapter. For many, that will be enough. For others, one or more other jobs are entailed. We refer you then to instructions in the later chapter.

It is advisable to have an assistant help you haul shingles to the roof, but if you must carry them alone, place them on a pad on your head for good balance and a clear view.

ASPHALT SHINGLING BASICS

You should know by now which facets of the job are necessary and what auxiliary jobs must be carried out. There may, however, be a problem with the sequence. You do not want to apply all the shingles and then find out you should have first put down the underlayment.

Work Sequence

Here is the sequential order of the different types of roofing work, for pitches from 2 to 21 inches. Steeper and flatter roofs are discussed later in this chapter, after basic procedures have been presented. Note the special instructions for roofs with pitches of 2 to 4 inches. Read the instructions on auxiliary and related jobs before tackling anything.

Structural deck repair or replacement. This step is necessary when your inspection shows rot or damage in the deck or framing. In many cases it calls for the services of a professional.

Eave and rake repair. Either of these, or both, may be necessary when old roofing is removed. Drip edges can be left on eaves if in good shape, but drip edges on rakes should be removed in any case, since they must be nailed over the underlayment. (See Chapter 7.) If underlayment is not used (as with wood shingles) rake drip edges can, of course, be left as is.

Metal flashing repair or replacement. This would not normally be necessary if roofing over an existing roof, but flashing should always at least be checked. On a simple roof, there may be no flashing except around the chimney.

Underlayment. New underlayment is essential for all jobs except wood shingles or when you are laying new asphalt strip shingles over old asphalt shingles.

Valley work. This will be needed

whenever the home is L-shaped, has gable-type dormers, or other areas where roof planes meet at an angle. Simple two-plane gable roofs have no valleys.

Laying field shingles. This is what most of us mean when we say "putting on a new roof". If you have a simple one- or two-plane gable roof, and are putting up new asphalt strip shingles over old ones, you can begin at this stage and ignore the earlier steps.

Ridge and hip shingles. This is the last stage. Ridge shingling is essential for every reroofing job. Hip shingling is similar, but necessary only on hip roofs.

Note that only the last two steps are essential in all cases. Some of the other steps may be needed as well, but we will discuss the more commonly needed jobs first.

NEW SHINGLES OVER OLD

If the old roofing material is of asphalt or fiberglass strip shingles, which are the most common type, and the new shingles are one of these, you can begin shingling immediately after preparation of the old surface.

This type of installation is considerably easier than most because you can use the old shingles to align the new ones. You do not need new underlayment, or new flashing, except at valleys or for needed repairs. Check the deck, rakes and eaves for possible repairs, and do this work beforehand. If you have an L-shaped home, gable dormers, or any other type of valley work, prepare these areas ahead of time (see Chapter 9 for valley details).

Assuming that all other preparations have been made, you can begin reroofing. One of the most difficult jobs will be getting the new shingles to the roof. If the dealer cannot put the material up there for you, be sure to carefully carry them up the ladder or scaffold, handling only as many at a time as you can carry safely.

The set of procedures presented here is designed to reduce unevenness due to new shingles on top of the butts of old shingles. The new horizontal nailing pattern falls 2 inches below the old roof's nailing pattern. The first instructions are for a 5-inch exposure for the new and old roofs, which is the most common situation.

Laying a 5-inch Exposure

If you must add new eaves flashing (see Chapter 7) snap chalklines on the eaves flashing. This will aid alignment of the new shingles until you reach the old courses, where the new shingles will butt against the old.

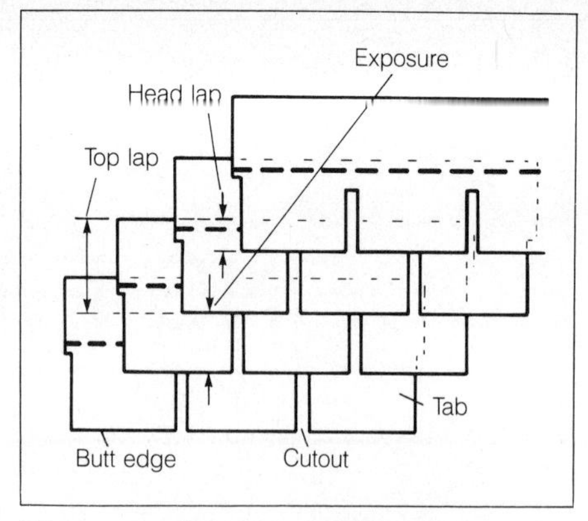

Whether the shingles are three-tab, two-tab, or have no tabs, the course above laps the course below at the top of the cutouts.

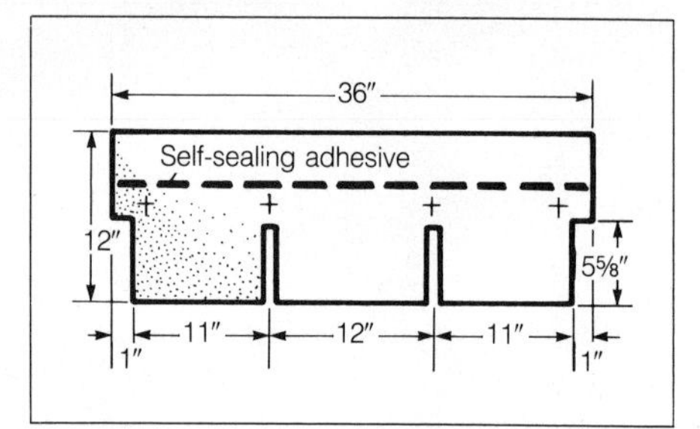

Four nails are always used. The center nails are directly above the cutouts for three tabs, or 11 inches from the outside nails.

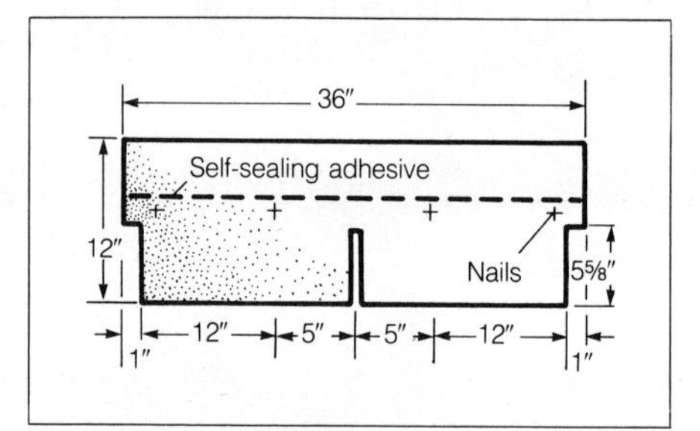

For two-tab shingles, four nails are still used. Nails are 5⅝ inches from the bottom, or halfway between sealant and cutouts.

Nailing the starter course. This type of strip should be nailed as for any other shingle strip, using four roofing nails. However, drive them in along the center of the strip. End nails should be one inch from the edges, and both center nails should be 12 inches in from the edges, as for field shingles.

How to cut shingles. Cut asphalt and fiberglass shingles from the back. Mark near the top and bottom edges of the shingle, from the front, using your knife. Then turn the shingle over and cut down between the marks. Save the larger cut-off pieces for use on the opposite rake.

Laying the starter course. To simplify alignment, and to avoid bulges from the old roofing, the edges of new shingles are laid flush with the bottoms of the old courses. Cut the starter course of shingles so its width equals the exposure of the old shingles, which in this case (and most cases) is 5 inches. Cut off the tabs and 2 inches or more from the top of the shingle with a utility knife or tin snips, until the strip measures five inches (or whatever the exposure is) from top to bottom. If using self-sealing shingles (as you should), position the sealant strip slightly above the eave edge.

You may choose to use a strip of mineral-surfaced roll roofing instead of the cut shingles. If so, the strip should be at least 7 inches wide. Nail the strip about

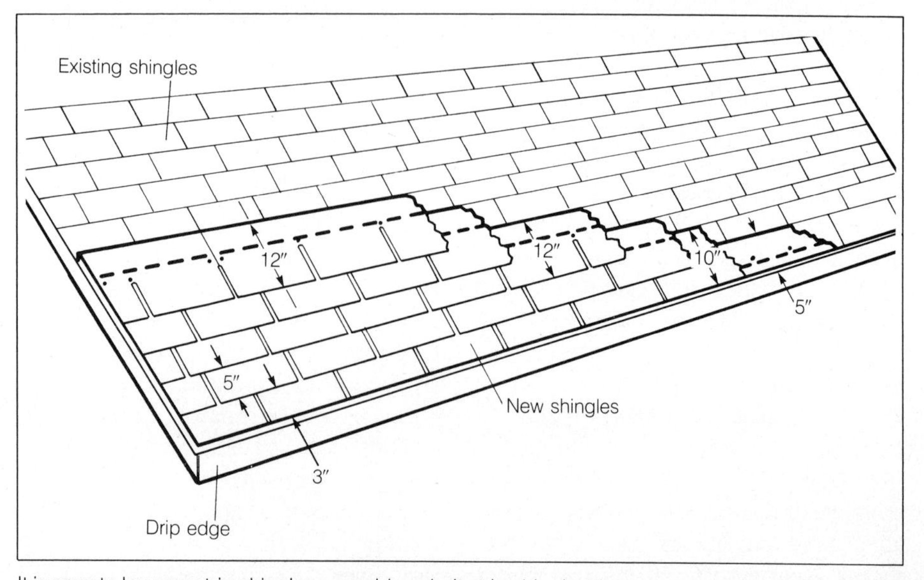

It is easy to lay new strip shingles over old asphalt strip shingles, because you can use old shingles for alignment. Note that the first course exposure is only 3 inches.

3 to 4 inches above the eave. Snap a chalkline as a guide. Lap ends of strips by at least 2 inches.

The starter strip should overhang the rakes or eaves by $\frac{1}{4}$ to $\frac{3}{8}$ inch, which should be flush to the existing eaves. If using self-sealing shingles, position the adhesive strip along the eaves. The shingles should overhang far enough so that water flows into the gutter, not into the exterior wall. If the existing eaves do not overhang this far, cut the starter strip to the width necessary for the overhang. However, do not overlap on the existing course above. Cut off 3 inches from the end of the starter strips first shingle, near the rake. This staggers the joints so that starter-strip shingle joints will be covered once the first course has been applied.

Laying the first course. Cut the shingles at the top, two inches down from the original edge. Fit the cut edge so that it is flush to the butt edge of the old shingle in the second course, down to the eave edge, covering the starter course. Nail these as for all field shingles, with four nails $5\frac{5}{8}$ inches from the bottom, or about halfway between the top of the tab cut-outs and the sealing material. Never drive nails into the sealant strip.

Laying the remaining courses. For the succeeding courses, full-depth shingles are used. The only cutting will be at the eaves. These shingles will all have 5-inch exposures. The top edges of the new shingles should meet the bottoms of the old courses. Butt each shingle against the old ones, but do not force them tightly. Proceed across and up the roof diagonally, until you reach the ridge. Cut at the eaves to match the old eave line exactly. Cut shingles to fit around obstructions. Apply roofing cement to bottoms of these loose edges and around joints.

Alignment patterns. On gable roofs, it is easiest to start in the far left-hand corner and work both to the right and up in a pyramid pattern. If you are left-handed, you will probably find it easier to start at the right.

For hip roofs, it is best to start at the center and work to both right and left, as well as up. Alignment should not be a problem when applying new asphalt over old, since the existing shingles serve as a guide.

You can produce a "starter strip" by cutting off the tabs of regular strip shingles. The amount cut off depends upon the exposure desired.

When reroofing the ridge, old ridge shingles must be removed before you can put up new ones. Here, a professional uses a roofer's hatchet; you probably will use a hammer.

Ridge and Hip Shingling

When reroofing over the old roof, remove the old hip and ridge shingles to obtain a level surface. This will also offer better fastener anchorage. Most manufacturers make matching hip and ridge shingles, and it is recommended that you use these. It is possible and a little cheaper to cut hip and ridge shingles from strip-shingle tabs, but they do not look as good. Never use metal ridge material with asphalt shingles; it may discolor due to corrosion. Vinyl can be used, but it will not match as well.

Bend each ridge shingle lengthwise so that it extends an equal distance down each side of the ridge. If you have to work

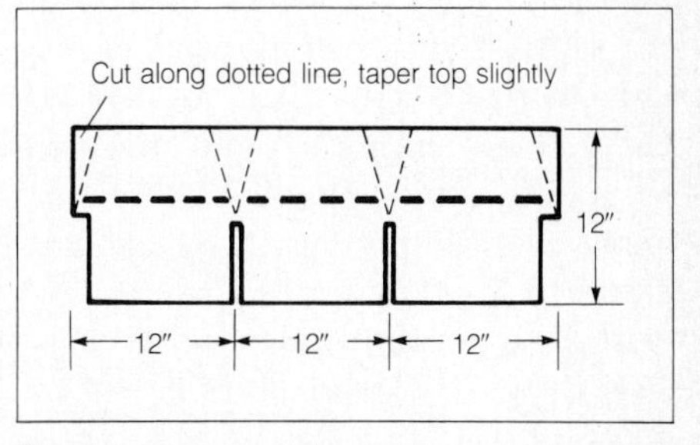

To create your own hip or ridge shingles, mark and cut to fit. Apply shingles from both roof sides; then finish the intersection.

in cold weather, you will have to warm each shingle before bending. About the only way to do this on a roof is with a blowtorch, which can be dangerous. The best advice is not to work in cold weather (for several reasons), but if you must, prebend the ridge shingles while still below, with a clothes iron. On warm days, lay ridge shingles out in the sun to warm.

Beginning at one end of the ridge, apply the shingles one at a time. Drive one roofing nail on each side of the ridge, one inch from each edge and 5½ inches from the exposed end. The exposure for both ridges and hips should be five inches.

Apply ridge shingles after field shingles. Use two nails, placed 1 inch from each side and 5½ inches from the exposed end.

Hip shingles are laid in the same manner, beginning at the bottom of the hip and working up. The first and last shingles will have to be cut to the shape needed. It will be easier to cut these from the mineralized surface of the shingle, unlike the field shingles. Tin snips may do a better job than the razor knife.

NEW ASPHALT STRIP SHINGLES OVER OTHER TYPES OF ASPHALT SHINGLES

New asphalt shingles can be laid over types of asphalt or fiberglass roofing—roll roofing, hexagonal, or individual lock-type—in much the same manner as previously described. The major difference is that you cannot use existing strip shingles as a guide. We must now discuss alignment and starting strips, the two elements not present when roofing over existing strip shingles.

Alignment

A poorly aligned roof will probably not leak any more than a properly aligned one, but it will not look right. There are people who can "eyeball" a straight line; this is possible with strip shingles, but most homeowner/roofers will do better with guidelines to help keep the shingles straight.

Chalklines. You need not line up each and every shingle separately; you should run a few chalklines for both horizontal and vertical alignment. Since strip shingles are three feet long each, vertical lines are recommended for every three shingles, or at nine-foot intervals, measuring out from one of the rakes. Snap a chalkline between these points from the ridge to the eaves.

Horizontal alignment is even more crucial than the vertical. To avoid the rakes ending up not quite equal, resulting in a crooked line of shingles at the ridge, measure down from the ridge on both sides every two feet. Snap a chalkline between the rakes at those points.

If you see that there will be a misalignment as you shingle toward those lines, gradually make any needed adjustment. Shingles have a tolerance of ⅛ inch on installation, so you can move them up and down as much as ⅛ inch on either side to achieve alignment.

Aligning in patterns. The usual way to align strip shingles is to lay each course so that the tabs fall halfway over the tabs below. When using this pattern, start the first regular course with a full shingle. Then the second course should have half a tab removed, the third course a full tab, and so on. This is the "6-inch method," which is described later along with the 4-inch, and the 5-inch methods.

"Random" shingling is not really random, although it is intended to appear so. It is achieved by making variations in the width of the end shingle cuts—but always in multiples of three. Leave at least three inches of tab at each rake, and locate cutout centerlines at least three inches from the cutouts in the courses above and below.

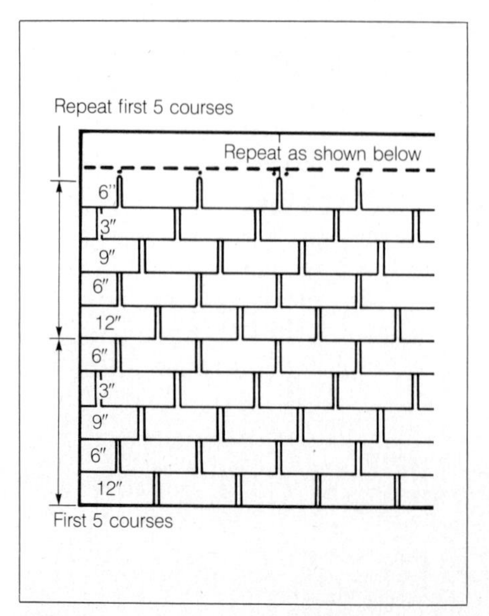

So-called random spacing is created by using the pattern shown. Shingles must be cut in multiples of three inches.

Laying the Shingles

The starter strip. An economical starter strip can be made by cutting off the tabs of regular field shingles, but it is simpler to buy a starter strip of 9-inch, 12-inch (or even wider) roll roofing of the same color. Whichever you use, nail it to the eaves so it projects ¼ to ⅜ inch over the eaves. Use 1½-inch to 2-inch roofing nails every 12 inches in the roll roofing, or four per shingle as in regular nailing. Avoid placing nails where they will be exposed by cutouts of the next course.

If you are using shingles as a starter strip, cut off three inches from the rake end of the first shingle. If using roll roofing, put a quarter-sized dab of roofing cement under the tabs of the next course of regular shingles. This is not required when using cut-off shingles, because the self-sealing strip is left intact.

Laying the first course. The first course of shingles, regardless of pattern, is begun with a full shingle laid directly on top of the starter strip and lined up with the edges at rake and eave.

Laying the remaining courses. Most roofers prefer to work in pyramid style, working over and up at the same time. To do this, lay two or three shingles of the first course, then lay one or two shingles in the second and third courses. Add another shingle to the first course (which is on top of the starter course) and another shingle to the second and third courses, and so forth. Proceed in that manner, following the chalklines.

For long or hip roofs. If you have a hip or exceptionally long roof, it is better to start at the center, or somewhere beyond the hip, with an artificial rake that pyramids in both directions over to the rakes. This permits better vertical alignment and avoids leaving less than three inches of shingle at either rake. (If necessary, move the centerline to the right or left six inches.) Be especially careful with alignment when working around dormers, chimneys, and other large obstructions.

STRIP SHINGLES ON BARE DECK

When roofing over a bare deck—whether on new construction, or when old shingles have been removed, or for plywood laid over the old shingles—the basic shingling technique is the same as just described. However, there are a few new elements involved.

Working Techniques

Underlayment. New underlayment must be laid when working on a bare deck, unless you have a flat or low slope of less than four inches of rise per foot. Even if you have removed old shingles carefully, and the underlayment looks satisfactory, take it off anyway. Underlayment is cheap and is easy to apply (instructions were given in Chapter 7) and there is no point in taking chances on old material.

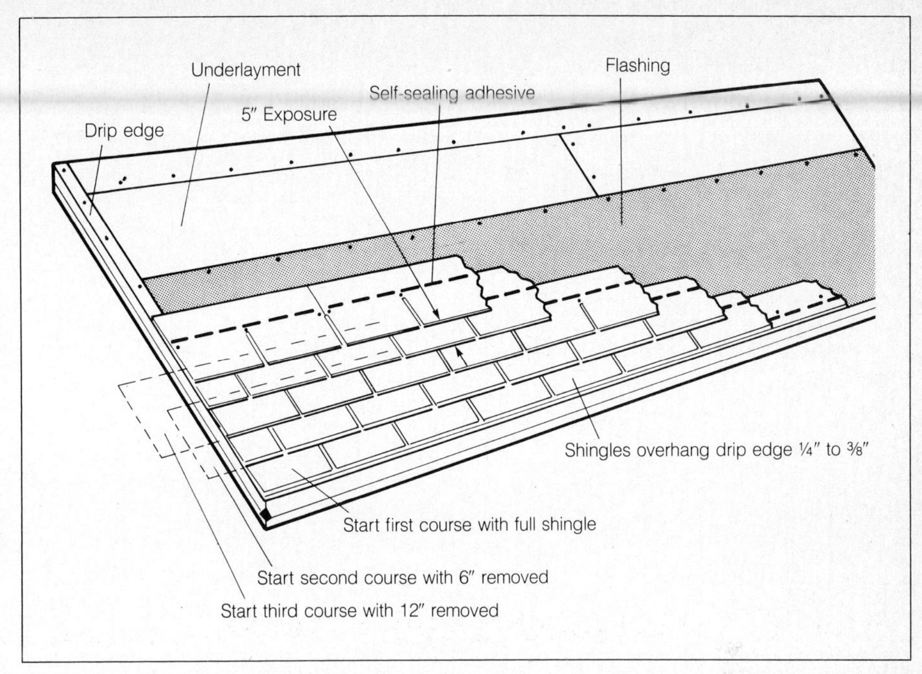

Lay the first course to cover the starter course. Begin at one side of the roof, and lay a full strip across the eave line. Cut 5 inches off the end of the next strip for the offset.

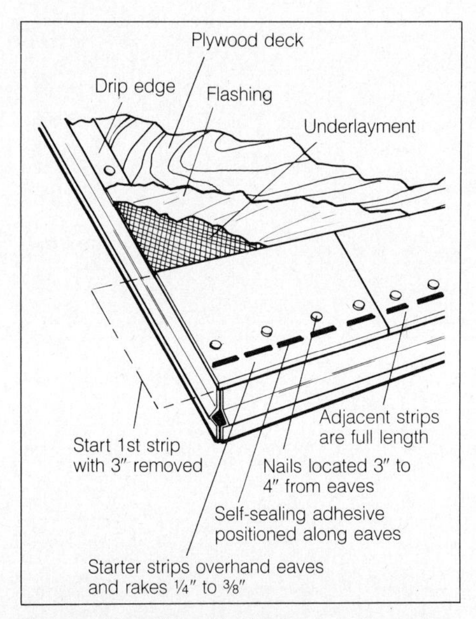

To use self-sealing shingles as a starter course, cut off the tabs and install with the cut edge at the eave line as shown.

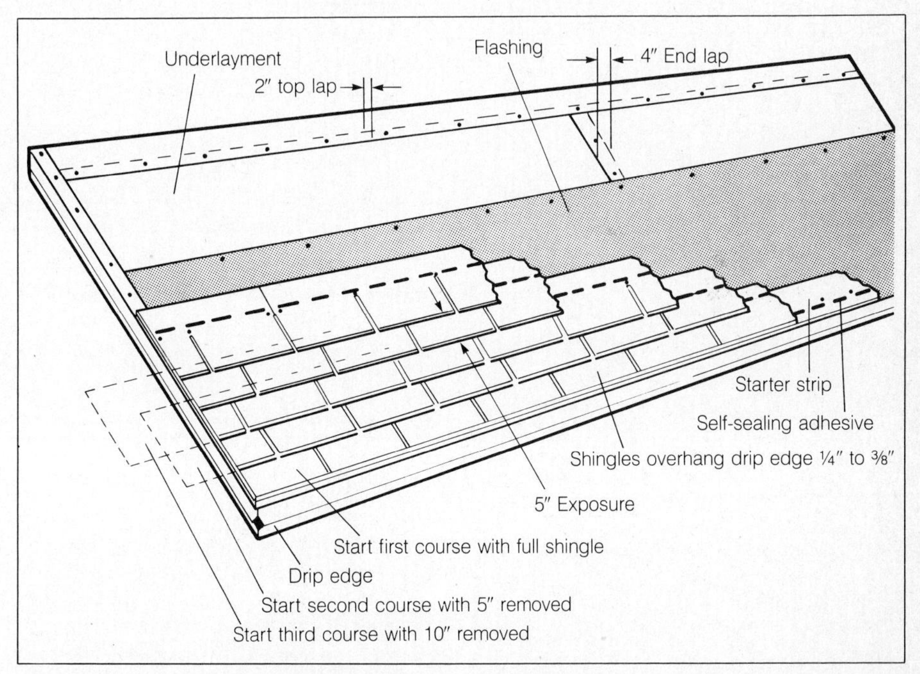

For a 6-inch offset, lay the first strip over the starter strip. Cut 6 inches from the second course strip. Begin laying the second course from the same side as the first.

Watertight surfaces. If the deck is a new one, first caulk all sheathing joints. Apply drip edges to eaves as described in Chapter 7. Remember that drip edges at rakes are applied after underlayment, but do not forget them once the underlayment is finished. On brand-new construction, the flashing work will be all brand new as well. See Chapter 9 for installation of new flashing. As you will note, flashings are applied at various stages of construction; it helps to make a checklist by sequence.

Slope requirements. On roofs with pitches of less than 1/6, or four inches of rise per foot, standard asphalt shingling methods cannot be used.

Shingling Low Slopes

Rise of two to four inches. Where the rise is from two inches to four inches, square-tab shingles approved for such use by Underwriters' Laboratories can be applied. They must be used with double underlayment and roofing cement. A special cemented eaves flashing strip also is required. To provide the eaves flashing, a continuous layer of roofing cement is applied at the rate of two gallons per 100 square feet under the entire first row of underlayment.

Laying the shingles. Low slopes require increased wind resistance, so use self-sealing shingles that have a factory-applied adhesive or else cement the tabs of regular shingles to the courses below. To secure the shingles that do not come

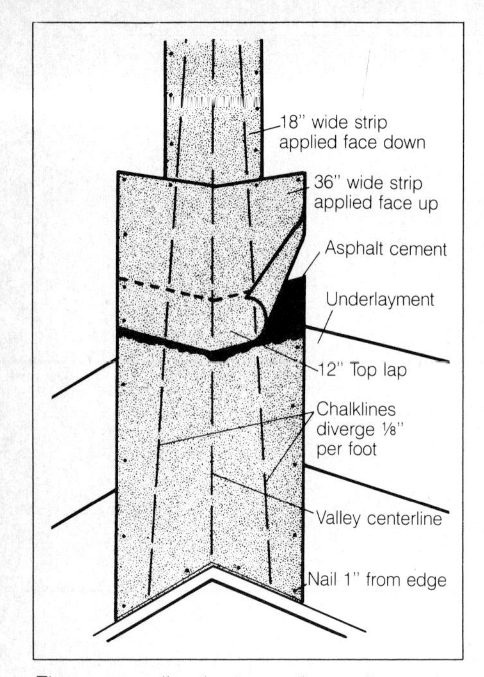

The open valley is the easiest valley to lay, although it is not as attractive as the woven or closed-cut methods.

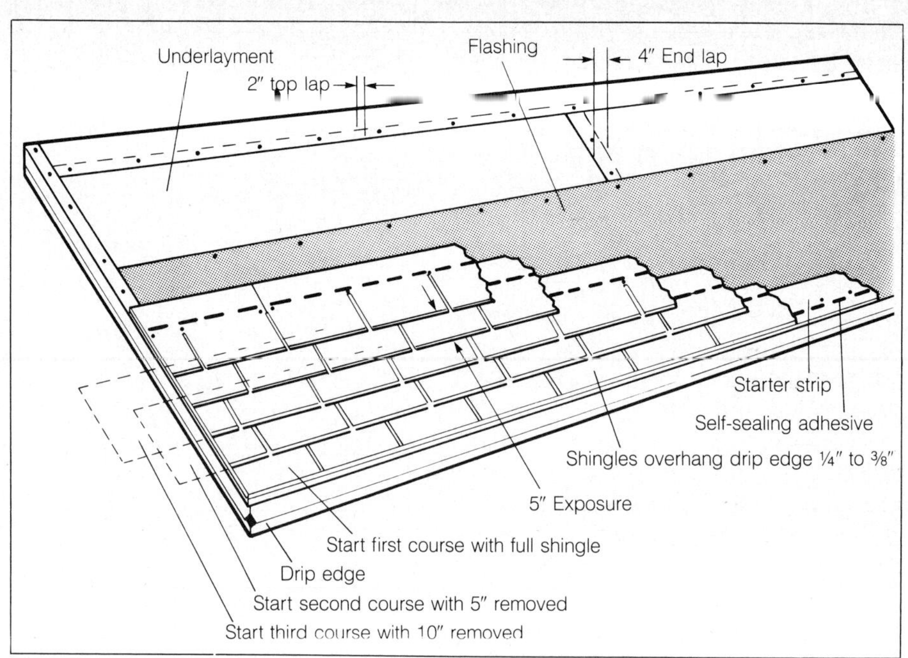

A four-inch offset gives a slightly different pattern to the roof, but the installation is the same, providing a 5 inch shingle exposure after the starter strip is covered.

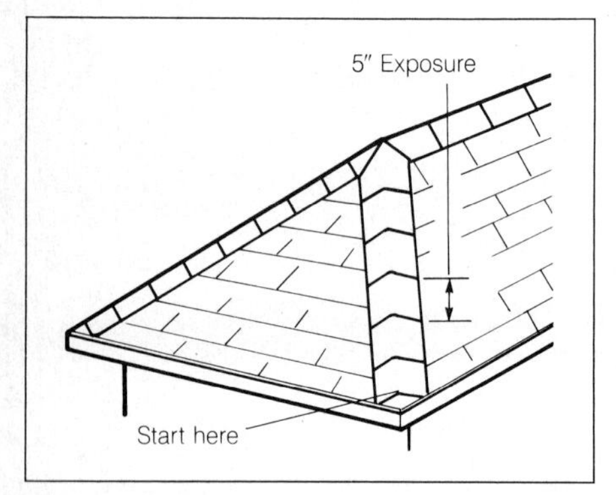

Hip shingles are laid in the same way as ridge shingles, but trim end shingles at eaves, edge, and ridge to fit.

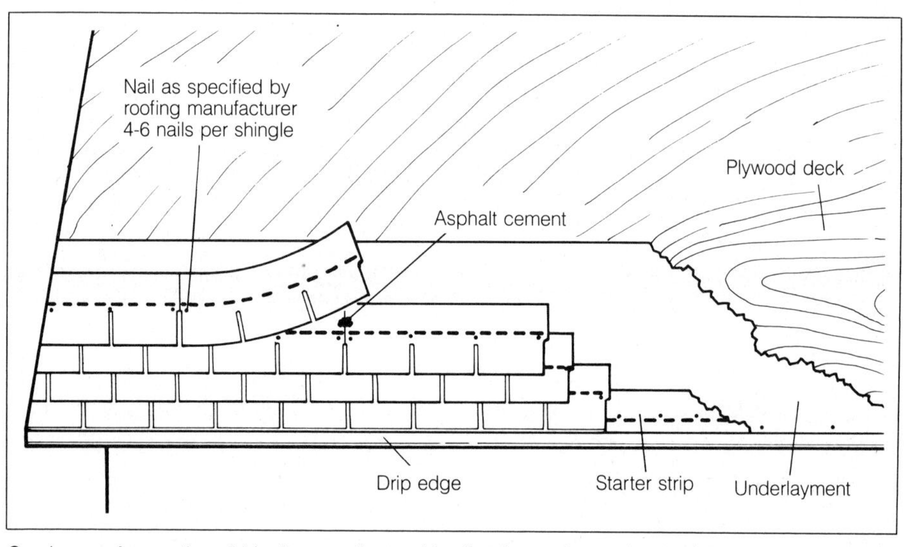

On slopes of more than 21 inches per foot, add adhesive under each tab of 3-tab shingle; 3 dabs under each tab of 2-tab shingle, use 3 spots for a 1-tab shingle.

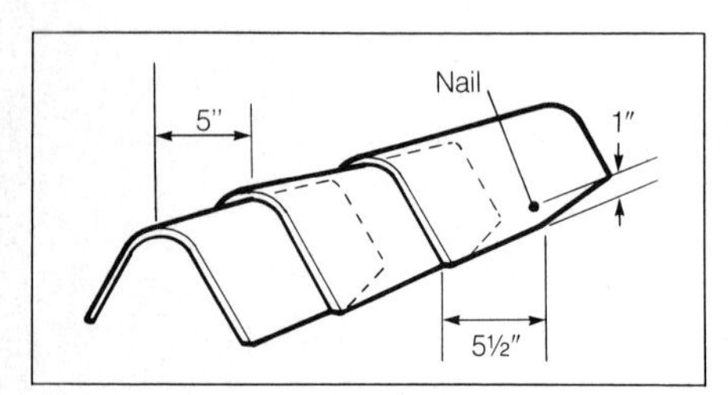

Shown are the preferred nailing (or staple) locations for shingles applied to hips and ridges. Use one fastener on each side.

with adhesive, place spots of asphalt plastic cement (about ¾ inch in diameter) under each tab. Press each tab onto the adhesive.

Rise of less than two inches. Pitches below two inches per foot must use roll roofing or some other nonasphalt roofing such as clay tile. Since roll roofing is not recommended for residential use, this process will not be described here. Flat roofs, as discussed previously, ordinarily have built-up roofing, which is a job for a professional roofer.

Shingling Steep Slopes

Pitches above 21 inches per foot. It is a rare conventional roof that exceeds a pitch of 60 degrees, or 21 inches per foot, but the steeper sections of a Mansard roof are often even more sharply angled. For Mansard roofs, you should be able to handle the steeper sections from a ladder. Roofs on some older homes may also exceed a 60-degree pitch. These very drastic slopes reduce the effectiveness of the factory-applied self-sealing adhesive, particularly in colder or shaded areas.

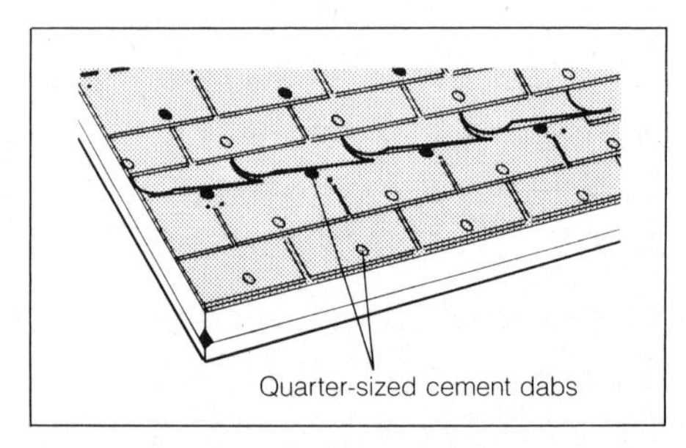

On steep areas of Mansard roofs, or where pitch exceeds 21 in. per ft., or ridges in high-wind regions, apply roofing cement.

For such steep pitches, extra sealant is necessary, even with self-sealing shingles. Put a dab of roofing cement, about the size of a quarter, under each tab of a three-tab shingle. If the shingle has only two tabs, put two dabs of adhesive under each tab. If shingles without cutouts are used, three spots of cement are required per 3-foot shingle. For these roofs, thorough ventilation is an absolute necessity.

Laying the Patterns

There are three common patterns of shingle installation: the 5-inch method, the 4-inch method, and the 6-inch method. The numbers 4, 5, and 6 refer to the additional number of inches cut off the first shingle in each course after the first course. The removal of these lengths of shingle result in the desired pattern. The cutting also helps stagger the cutouts, so they do not align from course to course. Although an alignment pattern using a cut-off of more than 6 inches is possible, never set up the shingle joints closer than 4 inches (in other words, do not try to use a 3-inch cutoff).

Do not throw away the portions you have cut off from the shingles. If a full tab is left, you can use it on the other side of the roof, or to create hip and ridge shingles.

For both free-tab shingles and roll roofing used for the starter strip, place a quarter-sized spot of asphalt plastic cement under each tab. Press the tabs firmly to embed them in the cement.

The first course. In all cases, the first course will start with a full shingle. Align shingle butts so they overhang the eaves by ¼ to ⅜ inch, completely covering the starter course. Snap a few vertical chalklines to help align the ends of the shingles. Use a horizontal chalkline to assure a perfectly straight first course.

For the proper exposure, lay all shingles so that the shingle butts will align with tops of the cutouts in the course below.

The 5-inch method. The first shingle in the second course will have 5 inches cut off. The first shingle in the third course has 10 inches removed. This progression continues through the seventh course, in which the first shingle has 30 inches removed. The second shingle in each course, and all adjacent shingles, are full length. In the eighth course, a full-length first shingle is again used. The cutting pattern is repeated, so that a full-length first shingle is used in courses 15, 22, and so on. The eighth course must begin the sequence anew because there would be only 1 inch left of the first shingle if 35 inches were removed.

The 6-inch method. The second course begins with a shingle from which 6 inches has been cut off. Each succeeding course has 6 more inches cut off. This means that the sixth course only has 6 inches left, and that the seventh course begins with a full-length shingle.

The 4-inch method. The second course starts with 4 inches cut from the end of the first shingle. For the third course, 8 inches will be removed. This continues through the ninth course, in which 32 inches are removed (4 inches are left in the shingle). The tenth course starts with a full-length shingle, and the sequence begins again at every ninth course.

Ridge shingles are less likely to resist high winds than field shingles are. Place a quarter-sized dab of roofing cement under each tab, and over nailheads.

Fastening Hints

The following tips will aid you in achieving a secure fastening job:

(1) use zinc-coated fasteners to prevent corrosion;
(2) avoid exposure of fasteners—shingles should cover fasteners in the course below;
(3) do not drive the fastener deep enough to break the shingle surface with the fastener's head;
(4) do not place fasteners in knot holes or deck cracks;
(5) always drive fasteners straight in;
(6) if the fastener does not penetrate properly, remove it, patch the deck, and drive another fastener nearby;
(7) do not nail into (or above) adhesive strip;
(8) keep cutouts and end joints at least 2 inches from nails in course below.

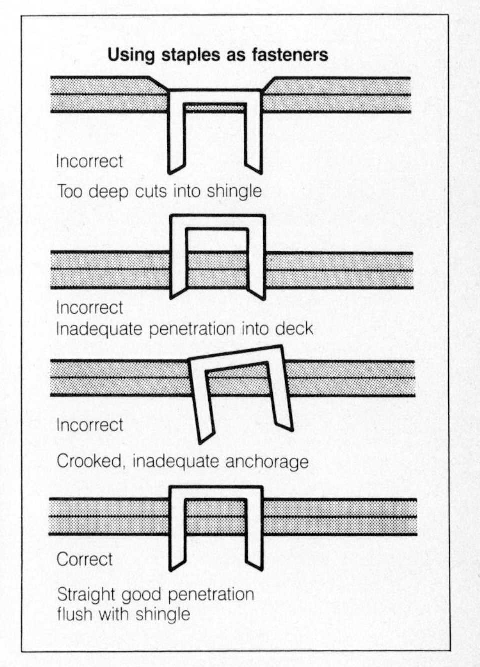

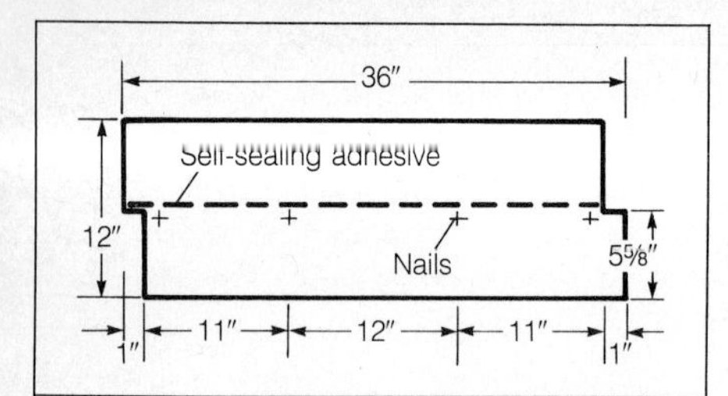

Shingles that are 36 inches long, but that have no cutouts, are nailed with four nails, similar to three-tab shingles.

To prevent buckling, start the nailing of each shingle from the end nearest the shingle just laid. Then work across. Once two nails are in place, do not attempt to adjust the shingle by shifting the end that is still free. This would cause distortion.

EXPOSURES FOR WOOD ROOFING*

Pitch	Lengths:	16"	18"	24"
No. 1 Blue label shingles				
3 in 12 to 4 in 12	Exposures:	3¾"	4¼"	5¾"
4 in 12 and steeper		5"	5½"	7½"
Shakes				
All pitches			8½"	10"

*See your dealer for exceptions.

Here, asphalt 3-tab strip shingles are being laid over wood shingles. The even surface is created by nailing beveled siding to the butts of the wood shingles.

USING CELOTEX R-I-F SHINGLES

When you purchase Celotex R-I-F (Root-It-Yourself) Kit, complete instructions are included. They are basically the same as for any asphalt strip shingles, except for the unusual pattern of the shingles. As shown, there are large cutout sections in which the tabs of each course line up directly above the cutout of the course above. Four nails are driven into each shingle, as for other types, with the placement in the narrow space between the self-sealant strip and the cutouts. It takes six courses to complete the pattern, which is then repeated.

Wood shakes need underlayment even when reroofing. For reroofing with wood shingles, no underlayment is used. Wood shingles require underlayment only for bare deck roofing.

WOOD SHINGLES AND SHAKES

Many of the same general principles for asphalt shingling apply also to wood shingles or shakes. Exposures vary with the pitch, and with the size of the shingle or shake. See the accompanying chart for proper exposures. Two nails are set into each shingle or shake, ¾ inch away from the edges and an inch above the exposure line (6 inches from the bottom for a 5-inch exposure, for example).

Underlayment over Bare Deck

Felt underlayment is applied differently for wood shingles and shakes than for asphalt shingles. Fifteen-pound roofing felt is interlaced with the courses. Much greater care is exercised in applying the felt than is the case for asphalt shingles, because the felt is used as a guide.

It is best if two people work on the underlayment. One person will roll out the felt along straight, premarked lines. The other worker will nail as the felt is unrolled. To fasten the felt, use 3d hot-dipped galvanized nails with a ⅞ head, nailing every ten feet, but only along the top edge of the felt.

The first felt course, along the eaves, should be 36 inches wide. Succeeding courses should be 18 inches wide. Cut 36-inch-wide rolls down the center if 18-inch-wide rolls are not available. The second felt course should be spaced so that the bottom edge is positioned 20 inches (twice the shingle or shake exposure) above the bottom or butt-line of the first course of shingles or shakes. Since shingles or shakes overlap the eave by 2 inches, deduct two inches from your measurement from the eave. With 24-inch shingles or shakes, for example, and a 10-inch exposure, the bottom edge of the second layer of felt is placed 18 inches up from the edge of the deck (2x10 inches=20 inches minus 2 inches.) The next strips are laid at the same distance as the exposure. In our example, with 10-inch

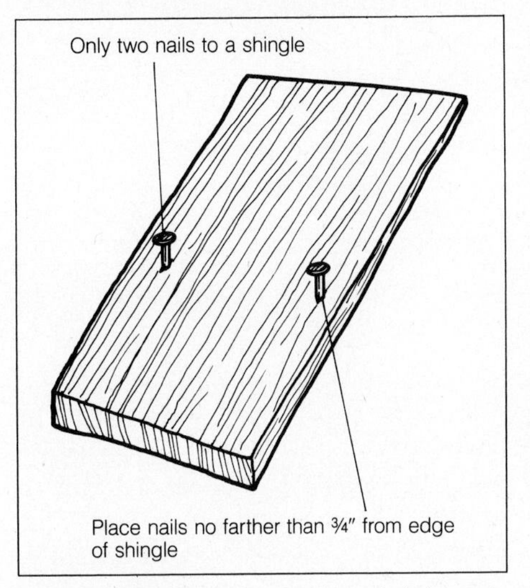

No matter what the size of the wood shake, only two nails are needed. Position them no farther than ¾ inch from each edge.

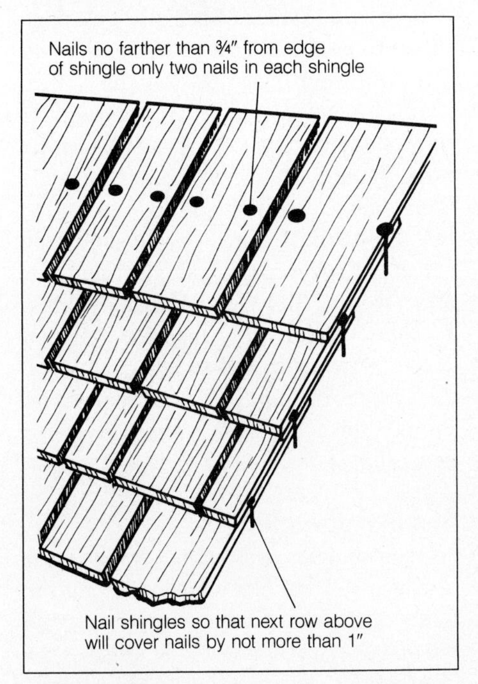

Nails should not be exposed. The course above reaches about 1 inch below nailheads in the course below.

Over a bare deck, wood shingles and shakes require roofing felt. The first strip is 18 inches wide; other strips are 36 inches wide.

The easiest pattern for a right-hand shingler is to work both up and to the right from the left-hand rake. Left-handers usually prefer to begin at the right rake.

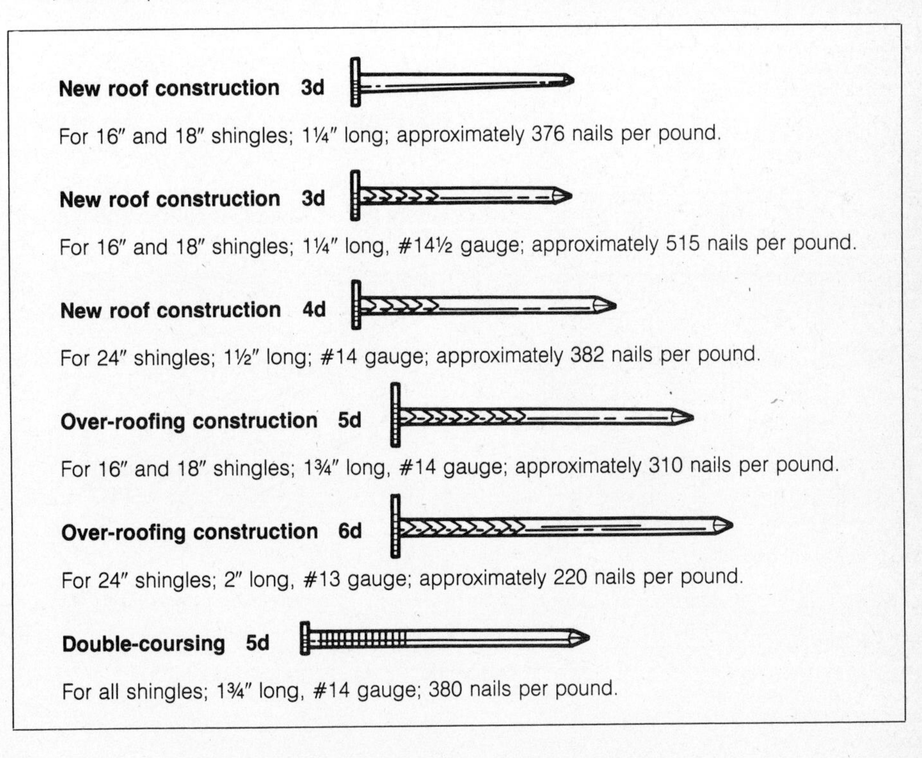

weather exposure, each succeeding course is laid 10 inches above the previous one for an 8-inch overlap

Laying the Wood Shingles

Over old shingles. If laying new wood shingles over old asphalt shingle roofing, cut away the old shingles from eaves and rakes (5½ inches from the edges) and replace with 1x6s, as described in Chapter 7. If applying over old wood shingles, also nail beveled siding to the butts of the shingles, as described in Chapter 7. To do this, remove the old ridge material (if you haven't already done so) and nail two pieces of beveled cedar siding, thin edge down, with the butts overlapping at the peak. If there are valleys, nail a strip of 1x6 lumber centered over the old valley flashing. Install new flashing as described in Chapter 9 and repair or replace any other flashing, as in Chapter 6.

Alignment. If the old shingle courses are straight, you can use them to help align the new ones. Otherwise, snap chalklines (as for asphalt shingles) with a horizontal line for each shingle course. Tack a long piece of straight 1x4, with a nail protruding slightly at each end, flush to the butt line of each course. Set the bottom of each shingle at the top of the 1x4. Move the 1x4 up for each course.

In reroofing, the sizes of the nails used when reroofing usually are 5d for shingles and 8d for shakes (2d bigger than for new roofing). Do not drive the nailheads into the shingles, but just to the surface.

Double the first course when applying over a bare deck. Start the first course so that the shingles project 1½ inches over the edge of the eaves. Double up the first course, by laying two courses on top of each other, allowing ¼ inch between shingle edges to allow for possible expansion. Stagger these two layers, with different spacing for the top course than the one below, to avoid continuous gaps between the shingles. The felt should cover the top four inches of the shingles.

Remaining courses. The second course is nailed at the proper exposure (see chart above) with joints offset at least 1½ inches from the joints of the previous course.

If you are using a roofer's hatchet, set the sliding gauge to the same distance as the exposure so that you can quickly check exposure dimensions as you work. As you start a new course, tack the 1x4 along the horizontal chalkline and butt the shingles edges against it for easy alignment. Use the head of the hatchet for nailing, and the blade to split the shingles when necessary. If you do not have a hatchet, a hammer and saw will serve the same purpose, although more awkwardly.

For 4 in 12 or steeper pitch, exposures are: 5 inches for 16-inch size; 5½ inches for 18-inch size, 7½ inches for 24-inch size.

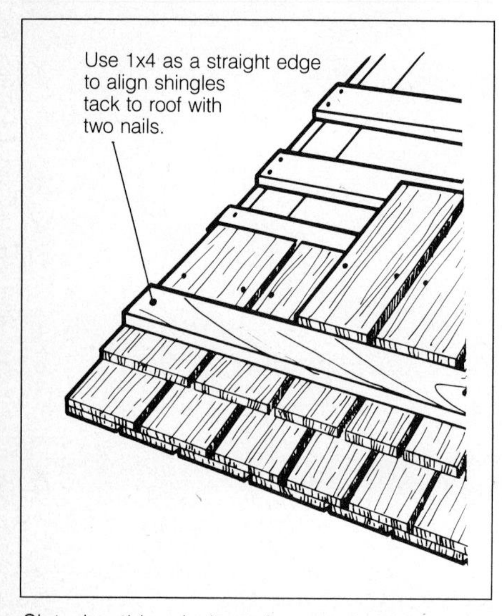

Slat sheathing is best for wood shingles use because the spaces between the boards provide necessary ventilation.

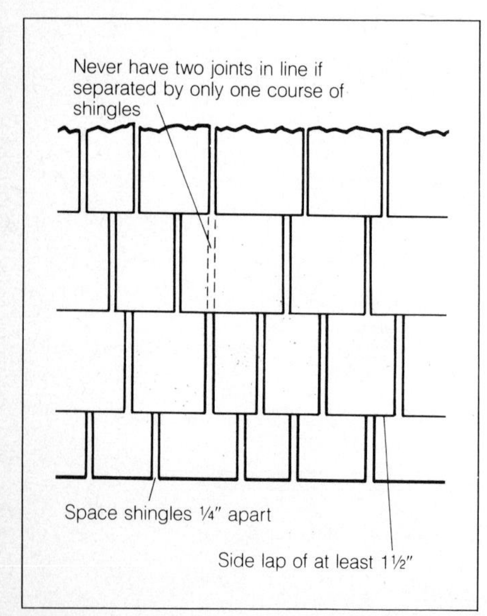

Space wood shingles ¼ inch apart. In succeeding courses, leave a gap of at least 1½ inches between joints of shingles.

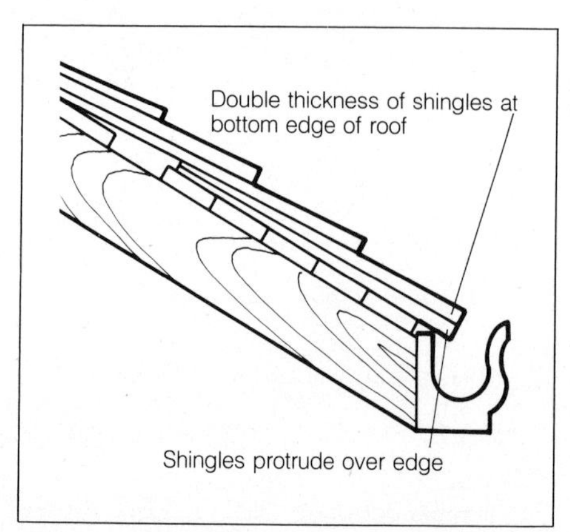

The doubled course of shingles at the edge of the roof gives extra eave protection and helps control water runoff into the gutters.

Slip each shingle under the felt course above it so that the top 4 inches of each shingle are covered by the felt.

Ridges and hips. Specially cut ridge shingles are the easiest to apply, but you can make your own "Boston Hip" ridge if you prefer. We advise getting special ridge shingles, since the two sides of the ridge should be beveled snugly against each other and must be sawed or planed to the proper angle. Each course should be nailed with alternately facing overlaps. In other words, first the right-hand shingle overlaps the left one, then the left one overlaps the right, and so on. Double the first ridge or hip shingle. Succeeding shingles take the same exposure as for field shingles.

Shake Application

As with shingles, start with a double layer of shakes for the first course. Stagger the joints of the shakes in the upper layer of the first course, relative to the joints of the layer below. The first course should project over the eaves by 1 to 2 inches.

Now place a strip of roofing felt over

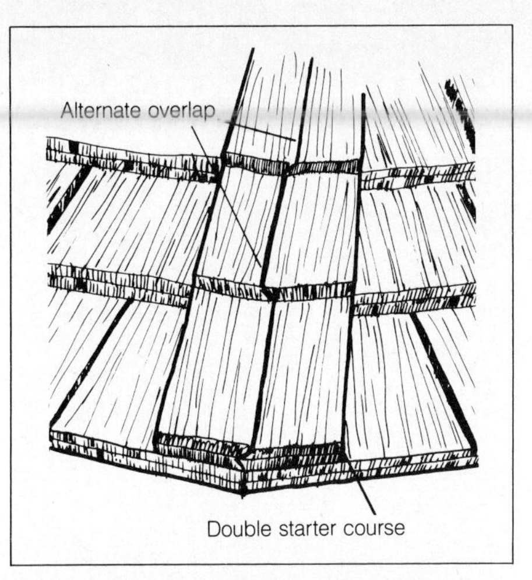

Double starting course at eaves of hips. Alternate overlap of shingles. Choose uniform 3-5 inch shingles.

Use metal valley material with interweaved strips of roofing felt for shakes and shingles. Shingles are trimmed to form a gutter.

the top edges of the first course of shakes. For the second course, lay the shakes as you would shingles. Check alignment with your gauge and a 1x4. Continue to alternate shake courses and to cover the top edges of the shakes with strips of roofing felt. (When professionals do this job, they lay all the roofing felt first, and then they slip each course of shingles underneath the rows of felt. This is a viable alternative, but it is harder to keep the correct alignment following this procedure.)

Saw shakes to fit at the ridge and at hips and at the ends of the courses. Cover the ridge and hips with hip and ridge shingles.

Important Construction Details

Walking over the roof. If you must walk on the new roof for any reason, use boards as a walk-way. Workers should always wear soft-soled shoes during shake application. They should never wear hob-nailed shoes. If these cautions are ignored, the result could be a leaking roof.

Buttlines at gutters. Because hand-split shakes come thicker than ordinary roofing materials, butt lines at eaves shed water runoff across a slightly wider area. This is especially true for roofs of hand-split shakes and resawn shakes. Use of wider guttering than is normally chosen will reduce splashing and runoff. For handsplit shakes ¾-inch thick or less, use 5-inch gutter widths. For shakes that are thicker, 6-inch widths are preferred. Use treated or painted wood gutters.

Use of felt interlay. Interlaced strips of roofing felt are recommended. These are of vital importance to the roof's future longevity. An underlayment of roofing felt, however, is of questionable value. It can adversely affect the roof's life expectancy in hot, humid climates because condensation may create a moist layer under the shakes. Do not use a roofing felt underlay instead of the recommended interlay.

Straight-split shakes. These are of equal thickness throughout. The smoother end ("froe" end) of the shakes should be at the top. This end, from which the shakes are split, allows for a tighter and more weather-resistant fit.

Finishing and Maintenance

Red cedar does not need any finishing or preservatives under ordinary weather conditions. It will weather naturally to a silver-gray color. Over a long period it will become almost black. In year-round warm, humid areas such as the American Southeast, or in any area below overhanging trees, a fungicide is desirable to inhibit moss, fungus and mildew. Some shingles and shakes may come pretreated. Apply toxic fungicides carefully, following manufacturer's instructions strictly.

Always clean wood roofs (and other types, as well) periodically to remove accumulated debris and to prevent moisture buildup. Use a stiff broom or brush to keep the joints clear between shingles, a vulnerable area in wood roofs.

Wood shingles that butt up to valleys, rakes, eaves, or the ridge, are cut to fit. When roofing over old asphalt shingles, use the old shingles as a guide.

Proper installation of shakes requires substantial overlap. The upper ends of the shakes fit under the strips of underlayment. The shakes are laid with staggered joints.

If existing flashing does not need replacement, be sure to caulk and seal completely after applying new shingles to the roof.

This roof needs flashing added to the brick wall. The homeowner is using a stop-gap seal of pressure-applied caulk to seal the joint.

In an open valley, the shingles are carefully trimmed at an angle and firmly embedded in asphalt roofing cement. When the trim is complete, it is advisable to seal the edges with asphalt caulk to be sure there are no crevices open to the weather.

Existing flashing requires periodic sealing to maintain complete protection against leaks. Asphalt roofing cement is used here.

APPLYING INDIVIDUAL SHINGLES

Individual shingles are manufactured in three basic types: hexagonal, giant and interlocking. Selection depends on many factors, including slope, wind resistance, coverage, esthetics and economics. Installation details vary according to the type of shingle used.

Preparation Procedure

Regardless of the type of individual shingle used on new construction, preparatory procedures are the same as those for applying strip shingles. Inspect the structure to determine if it is adequately ventilated. Cover the deck with an asphalt saturated felt underlayment. Strap horizontal and vertical chalk lines. Install drip edges, eaves flashings (if required) and valley flashings.

When individual shingles are used for reroofing, make the same preparations as those for reroofing with strip shingles. Evaluate the condition of the existing roof to determine whether the old material may remain in place. Perform remedial work on the deck if the old roofing is removed or on the existing roofing if it remains in place. Install flashings as required.

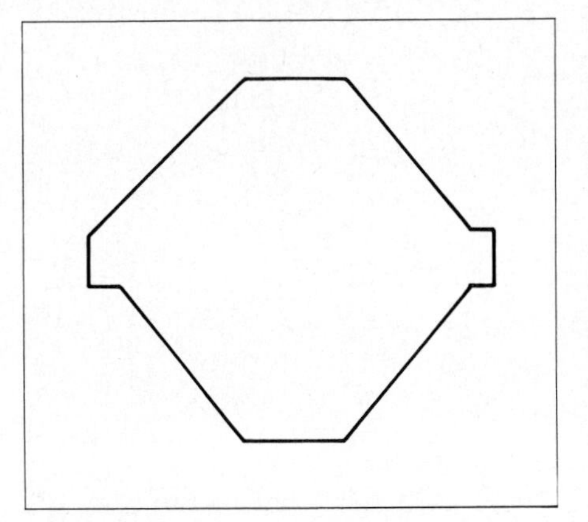

Hexagonal shingles are applied individually. Today, this style is not as commonly used as it was a few years ago.

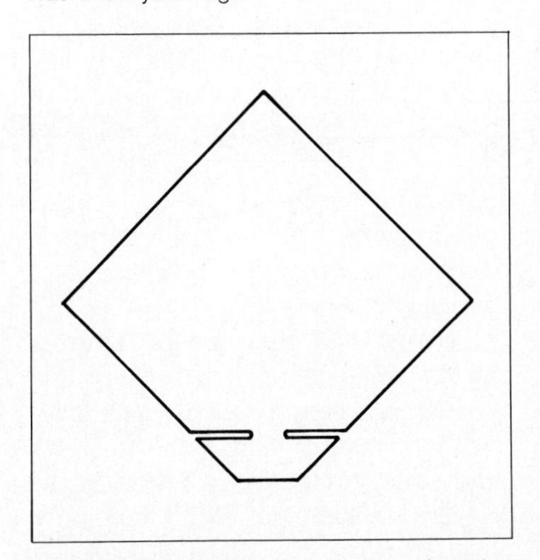

This lock-down style shingle is also referred to as a "hex" shingle even though it is a basically a diamond-shaped unit.

Laying Hexagonal (HEX) Shingles

Two types of hexagonal shingles are available: those that are locked together by a clip and those that have a built»-in locking tab. Both the clip-down and lock-down shingles are relatively lightweight and intended primarily for reroofing over old roofing. They may also be used at times for new construction. For either application, the slope should be 4 inches per foot or greater. Consult the roofing manufacturer for specific application instructions.

Laying Giant Shingles

This type of shingle may be used for new construction or reroofing depending on the method of application.

Dutch lap method. This method is intended primarily for reroofing over old roofing that provides a smooth surface and adequate anchorage for nailing. It may also be used to cover new decks where single coverage will provide the intended protection. For either application, the slope should be 4 inches per foot or greater. Consult the roofing manufacturer to specific application instructions.

American method. This method of application may be used for new construction or reroofing. In either case, the slope should be 4 inches per foot or greater. Consult the roofing manufacturer for specific application instructions.

Laying Interlocking Shingles

Interlocking shingles are manufactured with an integral locking device that provides increased wind resistance. The shingles may be used for reroofing over existing roofing on slopes recommended by the shingle manufacturer. They may also be used for new construction depending on whether single or double coverage is required. In general, single coverage interlocking shingles are not recommended for new construction. Check local building codes before applying them on new roofs.

The location of fasteners on the shingle is essential to proper performance of the mechanical interlock. For best results, follow the shingle manufacturer's specifications concerning fastener placement. Also follow the manufacturer's directions concerning application of the starter, first and succeeding courses.

Although interlocking shingles are self-aligning, they are flexible enough to allow limited movement for adjustment. Thus, it is especially important to snap horizontal and vertical chalk lines to keep the work in alignment.

The integral locking tab s are manufactured within close tolerances to ensure a definite space relationship between adjacent shingles. Be sure, therefore, to engage the locking devices carefully and correctly. The illustration shows two common locking devices used in interlocking shingles.

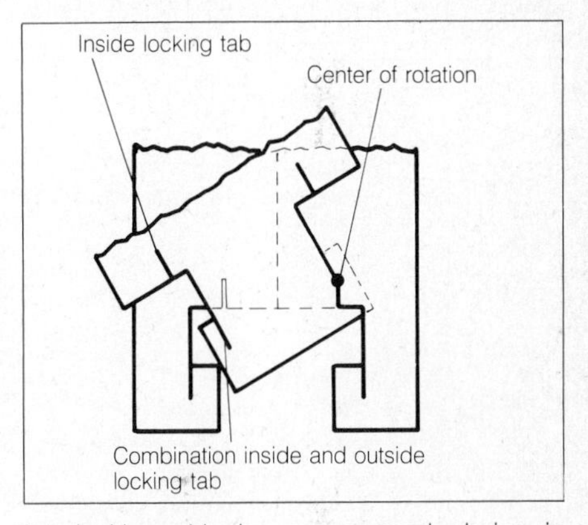

Interlocking shingles are a good choice in areas subject to high winds. This style interlocks with cut slots on short tabs.

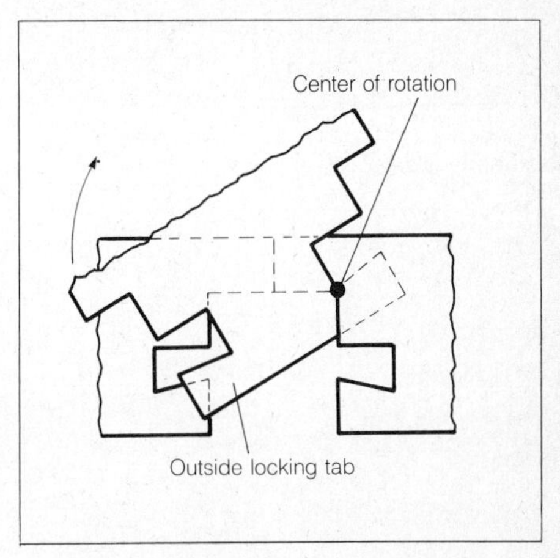

One version of the interlocking shingle uses long, uncut tabs to join individual shingles into a locked wind-resistant surface.

During installation, locking tabs on shingles along the rakes and eaves may have to be removed in part or entirely. To prevent wind damage, shingles that have their locking tabs removed should either be cemented down or nailed in place according to the manufacturer's recommendations.

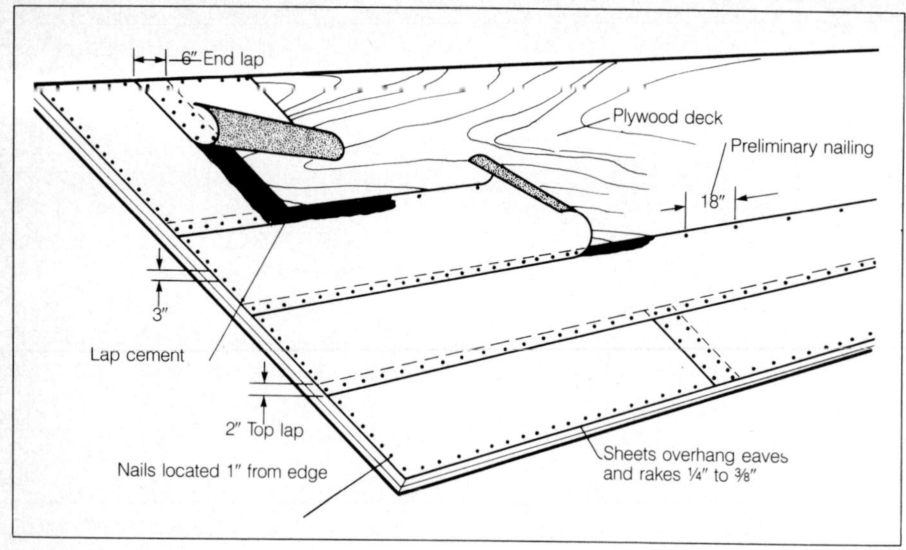

Exposed nail method of installation calls for roofing to be laid from the bottom to the ridge, parallel to the eaves. Cement bonds the 2 inch overlap that is also nailed down.

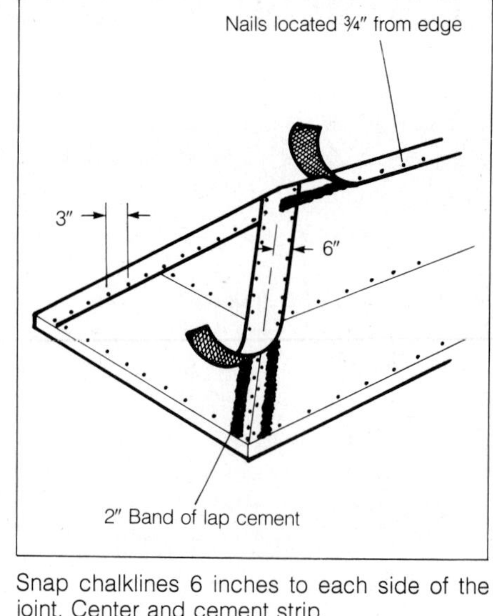

Snap chalklines 6 inches to each side of the joint. Center and cement strip.

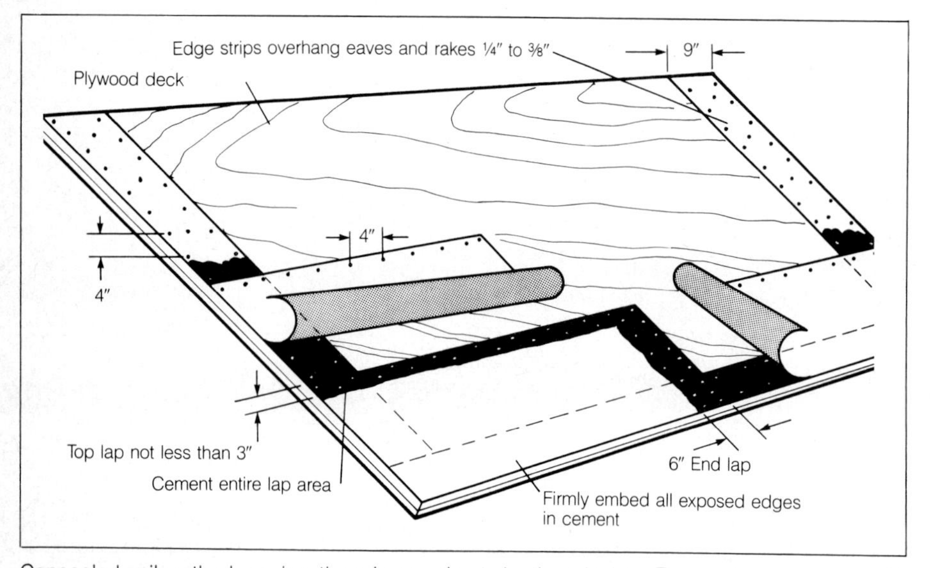

Concealed nail method requires the edge overlap to be three inches. The overlap is secured with cement. Starter strips run along the edges to accept cement and hold roofing down.

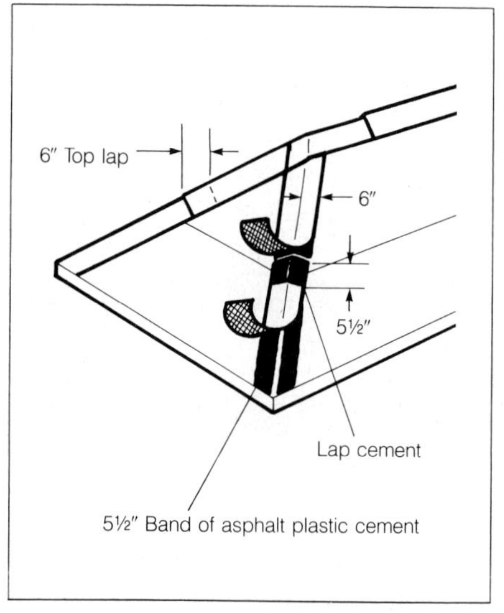

Ridges and hips are also nailed at the top under overlap and cemented down.

Lap cement
6" Top lap
3"
Nails located 1" from edge
2" End lap
Sheets overhang eaves and rakes ¼" to ⅜"

Application parallel to the rake requires a 2 inch lap on the long seam. Sections that overlap to complete a vertical course lap 6 inches. Laps are cemented and then nailed down.

APPLYING ROLL ROOFING

Single Coverage

Roll roofing is the best material for application to residential roofs with very low pitch. It has fewer seams and joints than shingles. Seams and laps are cemented and nailed, so there is less chance of leakage.

Roll roofing is laid parallel to the eaves or to the rakes, with exposed nails or concealed nails.

Exposed nail application. Whether applying parallel to eaves or rakes, the exposed nail method uses a 2-inch overlap of long edges and 6-inch overlap of short edges. Nails are 3 inches on center, staggered slightly to create two rows along all sides. The overlapped edge is cemented; then the upper edge is laid down and nailed.

Concealed nail application. Apply a 9 inch wide strip around the perimeter of the roof, nailing along both edges. Nail 4 inches on center, stagger nails so they are not opposite. Lay the starter course along the eave or the rake. Nail the edge to be lapped with two staggered rows of nails. Apply roofing cement at strip edges. Press the next roofing strip into the cement and roll flat. Cover nails with cement. Work from one side to the other to avoid blisters and wrinkles.

Double Coverage Roll Roofing

This method uses 19-inch selvage and 17-inch exposure, with a concealed nail technique. It is used for a nearly flat roof or one with a pitch of less than 2 inches per foot.

Cut a strip the width of the exposure. Nail it along the eave or rake. Apply the starter selvage so the edge overlaps the eave or rake by 1/4 to 3/8 inch. Nail selvage with 2 rows of nails; 4 3/4 inches below the upper selvage edge and 1 inch above the exposure. Apply roofing cement to the selvage and press exposure down with a roller. Do not use an excessive amount of cement. Roll from top to bottom of the strips and from one side to the other.

Lay each successive course so the exposure covers the selvage of the previous course. This creates a 2-inch, triple-thickness overlap at the exposure edge.

Exposed nail method. Apply a band of cement inside the chalkline. Center and nail a 12-inch-wide strip with nails 3 inches on center. Repeat for all hips; apply ridge last.

Hips and Ridges

Concealed nail method. Cut two or more strips, allowing 6-inch overlaps. Apply roofing cement between chalklines. Position the first strip as the bottom course. Nail with one nail on each side of the roof joint, 5 1/2 inches down from the upper edge. Roll out the strip to adhere. Repeat with next course. Overlap first course by 6 inches.

Double coverage roofing. Apply a selvage section to the hip bottom, nailing 1 inch from the edges. Position first course with exposure over the base layer. Nail with nails 4 inches on center, 1 inch from the edges. Apply roofing cement to the starter layer and press first course exposure into the cement. Repeat.

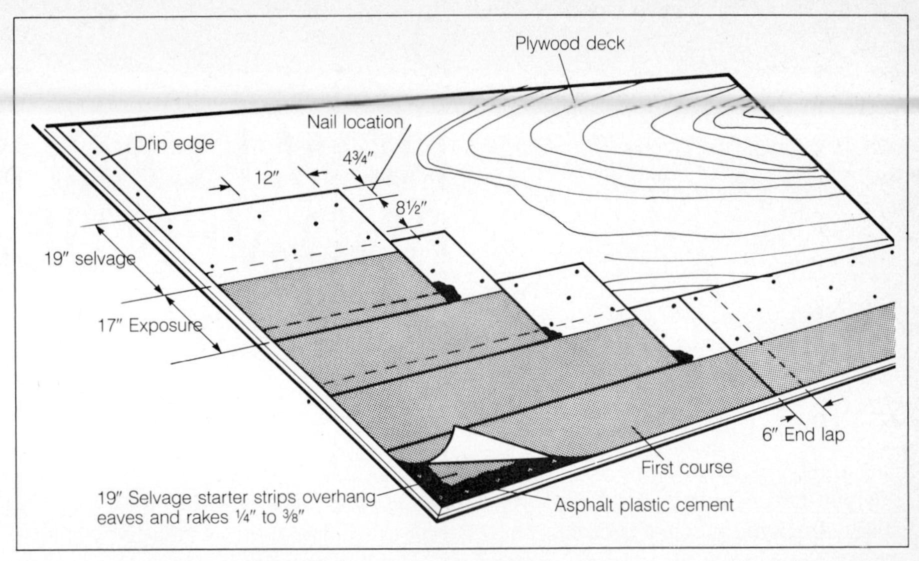

Double coverage roll roofing produces a triple thickness at each exposed edge. Selvage starter strip underlies the first course. Nail selvage down and cement exposure to it.

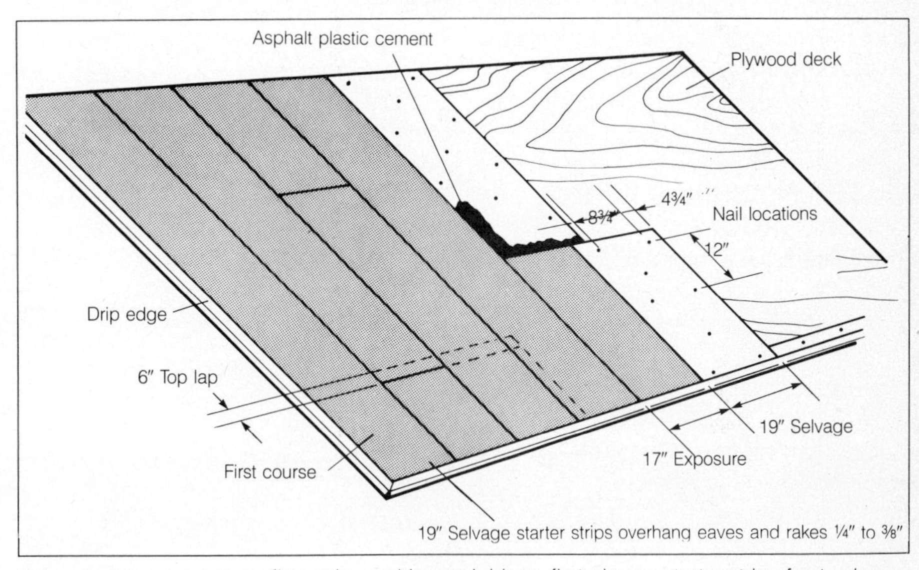

To lay double coverage roofing strips on hips and ridges, first place a starter strip of cut selvage. Apply cement to it and then apply the exposure of the first course.

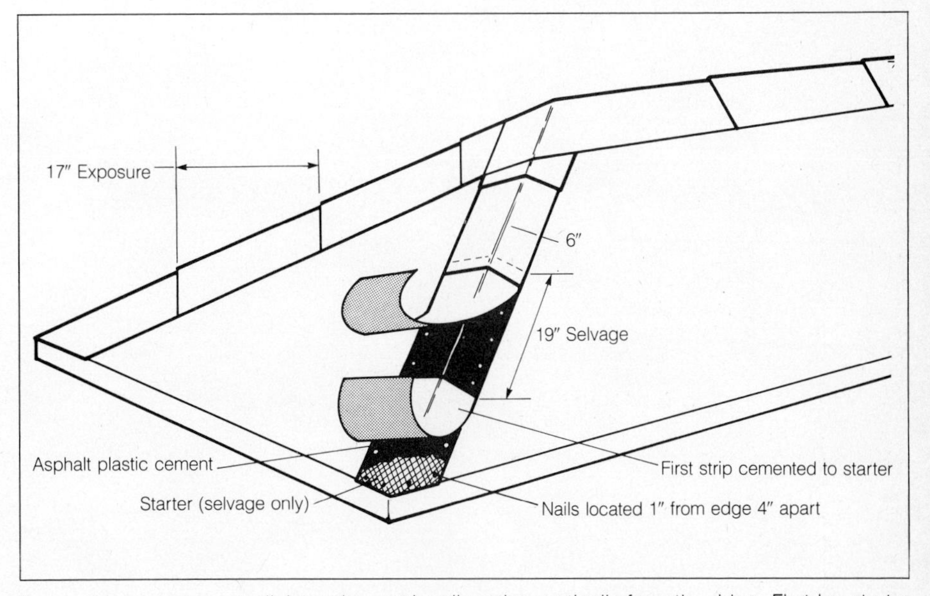

For double-coverage parallel to rake, apply roll rooting vertically from the ridge. First lay starter strips on rakes. Cover with roofing cement. Place full-width sheets.

9

NEW FLASHING AND VENTILATION

If you are fortunate enough to have a simple roof, you may not need this chapter. As discussed in the previous chapter, many roofs have no valleys, no gables or dormers, and no need for new flashing. If, however, your roof is more complex, or your inspection shows that new flashing is needed, refer to the appropriate section(s) below.

VALLEY FLASHINGS FOR ASPHALT ROOFS

There are several types of valley treatments, and because your roof has one type now does not necessarily preclude your selecting another when reroofing. In addition to metal valley flashing, there are three types of flashing that use asphalt products only. These include "woven", which has two interweaved layers of shingles and is least susceptible to wind damage; "closed-cut", which is similar; and "open", which leaves the flashing material exposed. Each method can be used for gable dormer valleys as well as for regular valleys.

This roof is in good condition, but the valley flashing needs some repair. Before painting or applying any sealing compound, be sure that the surface is clean and free of debris.

Installing a Woven Valley

If you are applying a woven valley over an open valley, the existing open valley must first be brought up to the same height as the old shingles by filling in with an extra layer of roll roofing. Use as few nails as possible—only enough to hold it down—and nail close to the edges. Do not overlap ends of this layer.

Step one: the roll roofing. The first step in installing a woven valley is to apply down the valley one strip of 36-inch, 50 lb. (or heavier) roll roofing. Using a minimal number of nails, set nails one inch in from the edges to hold the roofing in place. It will be easier to do this if you unroll the roofing in the sun before application. Where underlayment edges meet, overlap edges by 12 inches. Shape the ends to fit, using a utility knife or tin snips to cut the ends of the valley material

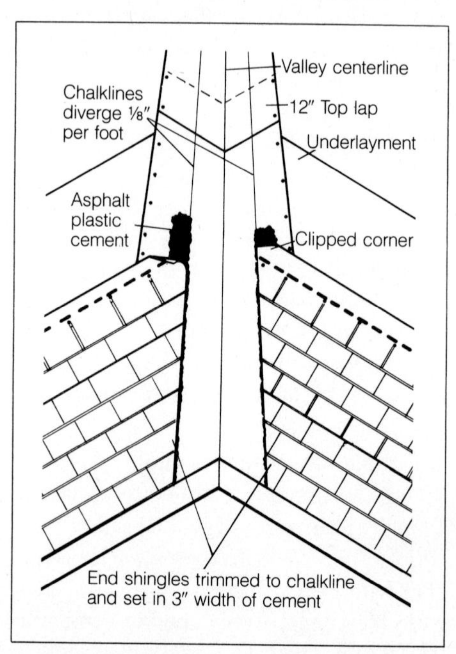

An open valley exposes the flashing material at the center. Shingles, cut at an angle, cover one-half the width of the flashing.

Shingles at an open valley are susceptible to wind damage. To lessen the possibility of their being pulled away, a generous amount of roofing cement is applied at the edges to hold shingles down.

to a V-shape in line with the edges of the shingles on both sides at the eaves.

Step two: laying the courses. When using a woven valley, it is best to lay the two intersecting roof planes at the same time. With a helper, this is easy, but it may prove cumbersome if you are working alone. In that case, apply shingles to a point approximately three feet from the valley, and finish the valley "weaving" afterward.

To make the woven valley, lay the first course along the eaves of one roof plane and into the valley. The last shingle will extend a minimum of 12 inches beyond the valley centerline. Press the shingles tightly into the valley but do not nail within six inches of the valley center. Adjust nail spacing to follow this rule, with one fastener at the top of each cutout and an extra nail at the top of the end of the strip.

Step three: weaving the courses. Now work on the same course from the other side of the valley, laying it across the valley over the course already applied and at least 12 inches beyond onto the roof surface. Then press into the valley and nail as for the previous course. Go on to the next course from the other side, nailing on top of the last course applied, and work back and forth from one side of the valley to the other. Remember to use two nails at the end of each shingle that crosses the valley.

Creating a Closed-cut Valley

Placing the underlayment. To create a closed-cut valley, bring up the level of any existing open valley, just as for a woven valley. Unless you are applying new asphalt strip shingles over old ones, cover the valleys with a 36-inch-wide strip of felt underlayment. Center the underlayment and hold it down with as few nails as possible, setting them 6 inches from the middle.

Laying the shingles. Run shingles along the eave of one side of the roof, as for woven valleys. Lay the remaining courses in the same way. They will extend across the valley and should end up on the adjoining roof plane. Press each shingle snugly into the valley. Use the same fastening methods as given previously, being sure that no nails are within 6 inches of the valley centerline. Place two fasteners at each side of every shingle that crosses the valley.

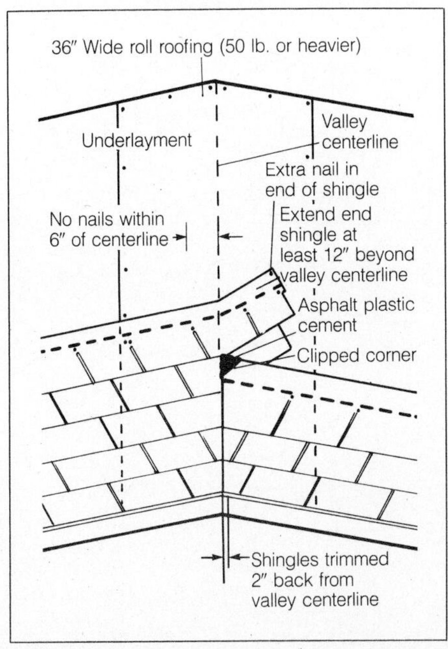

A closed valley is made by laying one shingle plane over the valley centerline and the other shingle plane up to that line.

On the second, adjoining roof plane, lay the shingles so they start along the eaves and cross the valley, overlapping onto the already applied shingles of the first plane.

Trimming and fitting. Trim the shingles so they are at least two inches away from the valley centerline. Snap a chalkline across the shingles and use it as a guide. At a 45-degree angle, trim one inch from each corner of each end shingle. This will help direct water into the valley. Then lay a three-inch-wide bed of roofing cement and embed the edges of each end shingle in it.

Laying an Open Valley

Step one: placing roll roofing. To make an open valley you will need two different sizes of 90-lb. roll roofing, one 18 inches wide and the other 26 inches wide. You can cut the 36-inch roll down the center to make your 18-inch wide roll.

Take the 18-inch roll and cut a strip to fit the valley, allowing 12 inches for overlaps of new strips. Seal the overlaps with an application of roofing cement on facing surfaces. Center the piece in the valley, with the mineralized side face down. Nail the strip one inch in from each side, using just enough nails to keep it smoothly in place. Nail one side of the strip, and then the other. The lower edge should be cut flush to the drip edge of the eaves.

The 36-inch roll is laid on top of the first, face up, and nailed in the same manner. If there are any overlaps, secure the overlapped portions (12 inches) with roofing cement.

Step two: snapping chalklines. Before applying the shingles, snap chalklines as placement guides. Snap two lines the length of the valley, one at each side of the centerline. Start it six inches apart at the top and let them diverge ⅛ inch per foot toward the eaves. (An eight-foot valley will be seven inches wide at the bottom; a 16-foot valley will be eight inches wide, and so on.)

Step three: shingle application. When placing the shingles at the valley, they should be cut cleanly at the chalkline. Use roofing cement under that part of the shingle that overlaps the valley material, and clip the upper edges as shown to avoid water buildup.

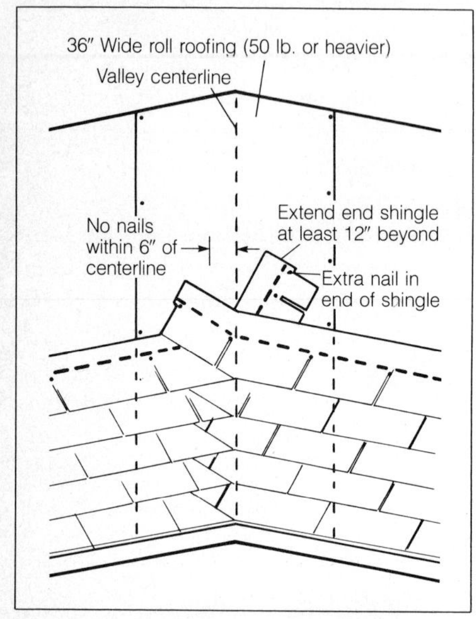

Woven valleys cover flashing completely and are seamless. This method is especially good in areas of high winds and storms.

Over existing roofing. Use 90-pound mineral-surfaced roll roofing to build up the valley flush to the level of the existing roofing. Then follow instructions as for new flashing. For open valley flashing, the new material will overlap the old shingles. For closed or woven valleys, the new shingles will cross above the valley filler strip.

Open Valleys on Gable Dormers

A little extra care is required when applying open valley flashing around a gable dormer. Here, after any underlayment is installed, apply the shingles on the main part of the roof just to a point above the lower end of the valley before placing valley flashing. Fit the last course close to the wall of the dormer and flash as necessary.

Step one: placement of 18-inch-wide roll roofing. At this point, the bottom 18-inch strip of roll roofing is applied similarly to that for other open valleys, described above. The bottom end, however, is cut so that it reaches ¼-inch farther than the dormer deck edges. Trim the lower edge of the strip that rests on the main deck so that it protrudes two inches (or more) beyond the point where the two decks join. If the old flashing has been left in place, use it as a guide. Place the upper end of the strip so that the section on the main roof reaches 18 inches higher than the point at which the dormer meets the roof. Cut the dormer strip off at the roof ridge.

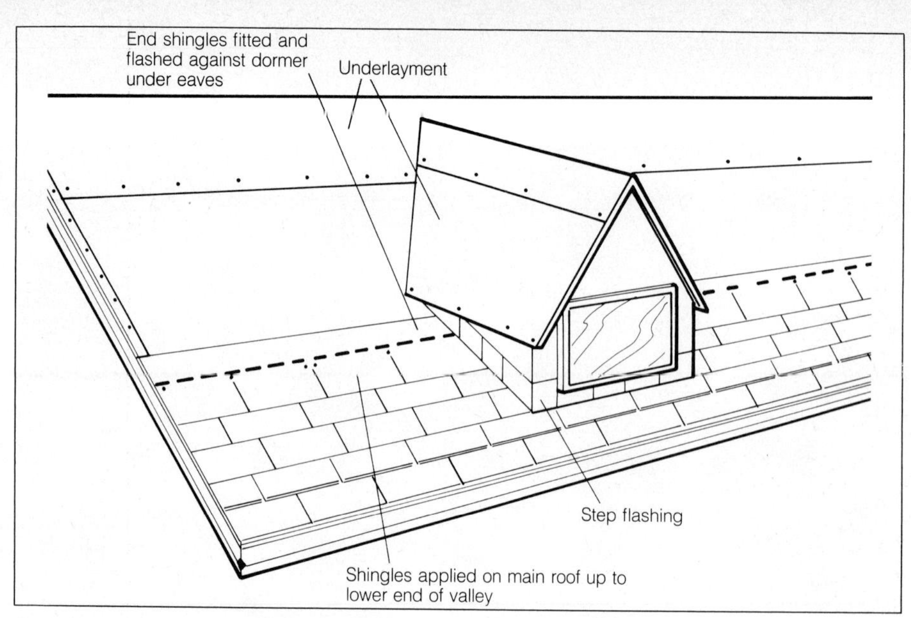

Gable dormer walls require step flashing installed from front to back so that water will flow over and not into the joints. Flashing fits against the roof and under the shingles.

Repeat this process for the other side of the dormer. Extend the top portion of this second side upward and over the first piece of flashing, to overlap by 12 inches. Use nails and cement on the overlap.

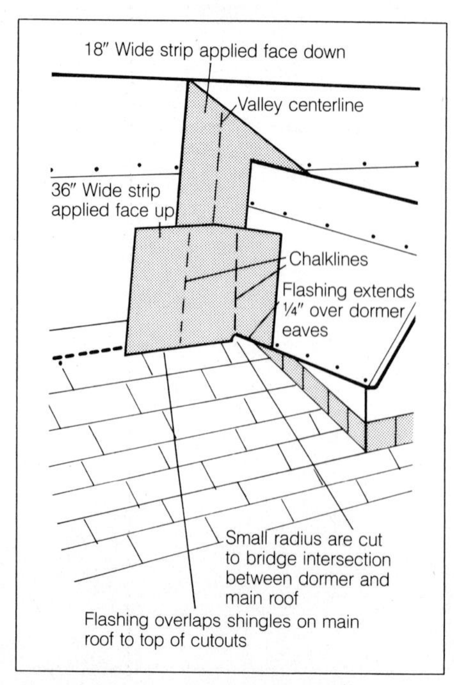

Gable dormer roof valleys are covered with flashing in layers for good coverage of the joint between gable roof and main roof.

Extend the flashing on the second side of the dormer so that it extends over the ridge. Again, nail and cement the overlap.

Step two: placement of 36-inch-wide roll roofing. The 36-inch flashing strips of roll roofing is then cut on the dormer side so that it matches the eave end of the 18-inch strip. On the side that will overlap the main deck, cut the lower end to overlap the closest course of shingles, along the same overlap line as for an ordinary shingle. When using the usual three-tab strip shingles, cut the roll roofing to the top of the cutout of the shingle course below.

Adjust the top layer of flashing so it lies smoothly in the valley joint. Nail it over the 18-inch layer just as for a regular open

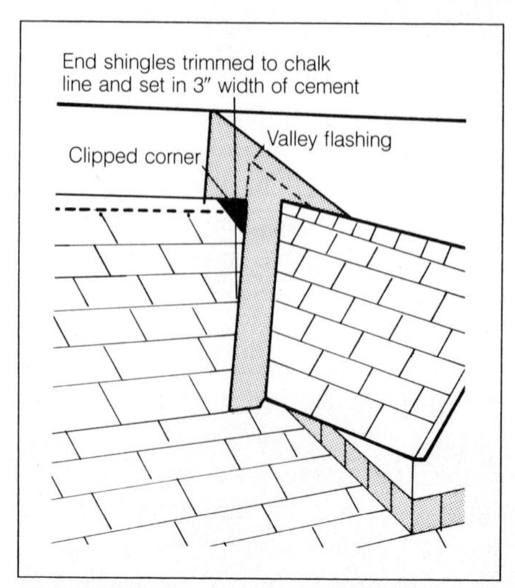

A completed valley sends runoff away from the dormer onto the main roof. Shingles are cemented to prevent water erosion and leaks.

valley. Cut off the top edge so that it follows the horizontal line of the dormer ridge. The lower edge of the flashing should be cut along a small-radius arc to bridge the intersection of dormer and the roof plane, leaving 1/4 inch projecting beyond the dormer eaves. When finished, the valley flashing should form a small canopy over the joint.

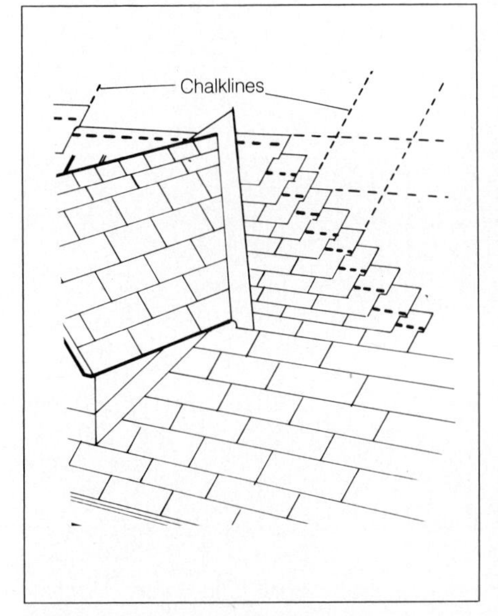

After shingling the dormer roof, lay main roof shingles up to the dormer peak. Adjust overlap so courses align at the peak.

Step three: shingle application. Continue to lay shingles, cutting the end shingle of each course so it aligns with the chalklines. Then trim the upper corner of each end shingle so it can be embedded in a strip of roofing cement. This seals it to the flashing. Continue valley shingling, as above.

To ensure alignment at the top of the gable, lay a few trial courses as you approach the peak. That way, you can gradually make any necessary adjustments over several courses. As noted, asphalt shingles have a 1/8 inch tolerance, so that you can adjust up to 1/8 inch per course, that's all.

Metal Flashing on an Asphalt Roof

Although roll roofing flashings are common when reroofing, metal flashing material is more often used for new decks. Sheet metal, crimped in the center and at the edges, is used instead of roll roofing. The flashing should extend 10 inches on each side of the valley, which means you need 20-inch-wide metal.

After the flashings are in place, coat the edges with a heavy dose of roofing cement before laying the shingles next to the valley. Clip shingle edges (as discussed above for "valleys") and coat the underside of each edge shingle with more roofing cement. Keep nails at least six inches away from the center of the valley.

FLASHING AGAINST A WALL

Side Wall Against New Deck, Asphalt Shingles

When any roof plane butts against a vertical wall—on a shed dormer, for example—flashing should be applied between the roof and the wall. Metal flashing is used on new roofs, but when reroofing, you also need roll roofing.

Using flashing shingles. If you are flashing over a new deck, ask your dealer for "flashing shingles". These are pieces of thin weatherproof metal that come ten inches long and two inches wider than the usual five-inch shingle exposure (seven inches). One end of this type of flashing will have to be covered on the sidewall by the siding, so pry up the siding at least five inches back if the siding is already in place. (This is one of the reasons roofing is applied before siding if you are building a new house, addition or dormer.) This type of application is called "step flashing."

Installing the first step flashing. First, run 3 or 4 inches of the underlayment against the wall, but do not nail it to the wall. Take the first flashing shingle and bend it in half, into two 5-inch legs that form a 90-degree angle. Place it against the wall so that it overlaps the end of the starter strip and so that the tab of the end shingle of the first course will cover it entirely. Use two nails to secure the horizontal section of the roof; do not nail the other end to the wall. Now apply the first shingle course until it reaches the wall.

Adding succeeding units. Place the second step-flashing shingle over the shingle at the end of the first course, laying it 5 inches up from the shingle butt; the tab of the shingle at the end of the second course will completely cover the step flashing. Then nail down the step flashing. Lay the second course of shingles. Repeat the sequence up to the top of the intersection. Since the metal strip is 7 inches wide, and each roof shingle has a

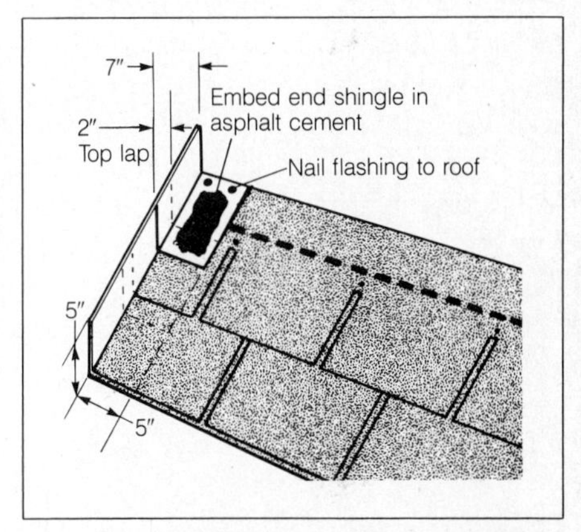

The correct placement of step flashing is shown. Overlap on both the wall and the roof to prevent water backup and leaks.

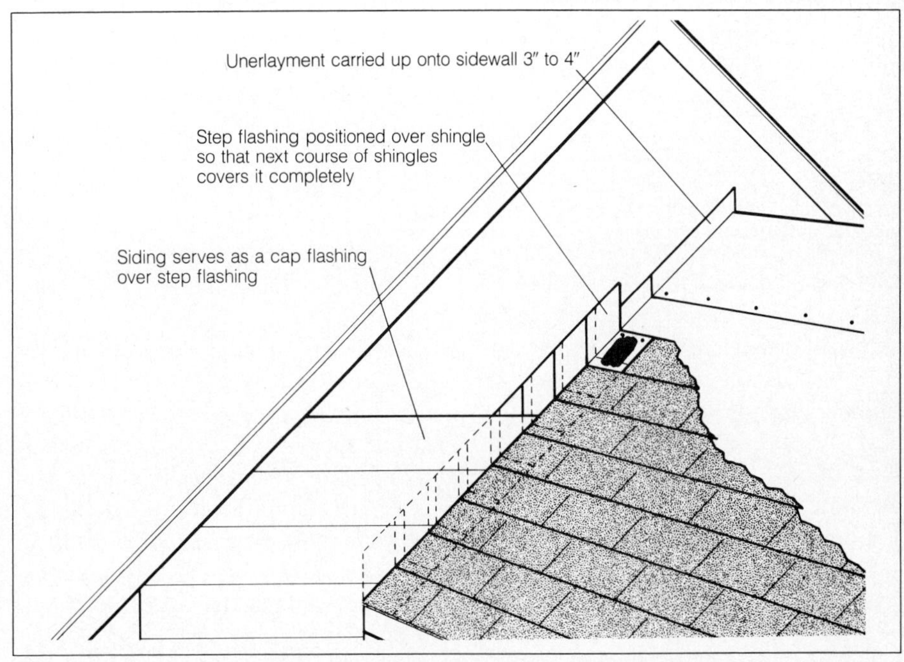

Step flashing overlaps underlayment, shingles overlap step flashing, and gable siding overlaps upper sections of step flashing. Properly applied, flashing will resist water and wind.

5-inch exposure, each piece of flashing overlaps by 2 inches the one in the course below.

Finishing the work. Siding should be brought down over the vertical flashing. Nail the siding over the flashing, making sure that there is a wide enough gap between the siding and the roof so that you can paint the ends of the siding boards. The siding in this case also serves as "cap flashing" (below).

Against Vertical Front Wall

Lay shingles up to the bottom of the vertical wall. If necessary, adjust the exposures in the last two courses so the last course, which will be trimmed, ends up at least 8 inches wide. Add a continuous strip of metal flashing on top of this last course, embedding it in roofing cement. Nail it to the roof. The strip of metal flashing should be 26 gauge. Bend it so it will reach a minimum of 5 inches onto the vertical wall. At least 4 more inches of the strip should extend over the last course of shingles. Lay a trimmed row of shingles on top of the metal flashing strip, as shown.

Lay a trimmed layer of shingles over metal flashing at a vertical front wall. Embed them in roofing cement; butt them to a chalkline snapped on the last row of full shingles.

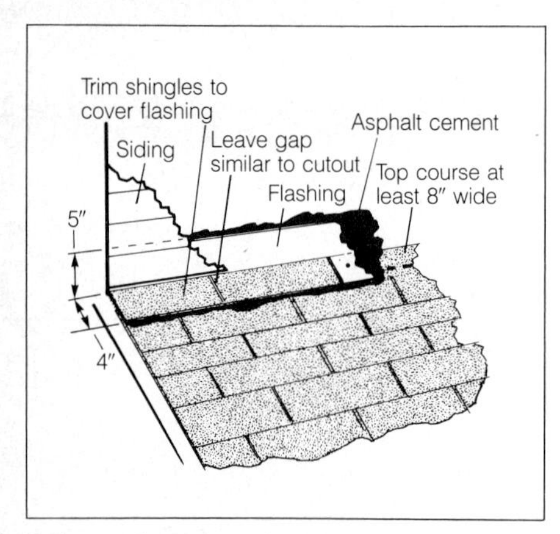

If the roof meets a vertical wall, the joint is covered with flashing and a line of cut shingles is firmly cemented over flashing.

Bring the siding down as for the sidewall. Where a vertical front wall meets a sidewall, trim the flashing to let it reach 7 inches around the corner. Then place step flashing along the sidewall.

Asphalt Shingles Over Wood Shingles

Use a strip of 50 lb. smooth roll roofing, 6 to 8-inches wide, laying it across the wood shingles that butt against the wall. Nail along each side of the strip, spacing nails every 4 inches. Apply a thin layer of roofing cement and securely embed the end shingle of each course in the cement. At the joint between the new shingles and the sidewall, insert a bead of caulking cement. When working over an existing roof and sidewall, roll roofing is added once the old roofing has been leveled.

CHIMNEY FLASHING

Chimney flashing application is complicated. If building a new home or addition, which requires the use of virgin flashing around a new chimney, and if this is one of your first roofing ventures, have a mason do the flashing (as well as the chimney itself). Although roll roofing can be used for chimney flashing, metal flashings last longer and are more commonly used.

Fortunately, these flashings can be left in place when reroofing, as long as they are not badly deteriorated. Carry out any flashing repairs before reroofing. If any portion of the flashing needs replacement, install the new sections as explained below. In most cases, reroofing calls for embedding edge shingles in roofing cement and caulking the cracks with the same material.

Location. Chimney flashing varies with the location of the chimney, which can be at the rake on the gable end of the house, on the sloping part of the roof, or at the ridge. Both base and cap (or counter) flashing are needed. Although not every possible situation can be covered here, there are a few basic applications that apply to nearly every situation.

Using Base Flashings

Base flashing is roll roofing that attaches to the roof deck under the shingles and is bent up and secured to the chimney with one or two masonry nails driven into the mortar between the bricks. You may wish to hire a professional to repair damage or deterioration found in the base flashing. It requires the removal of the counter flashing to provide access. Fortunately, base flashing rarely needs repair.

Substituting step flashing for base flashing on vertical masonry wall or chimney. This variation on the step Pflashing method (discussed earlier) requires placement of an 8x22-inch piece of flashing material over the end tab of each shingle course. The lower edge is kept one inch away from the exposed edge of one of the shingles in the strip. The upper edge of the flashing bends up against the masonry.

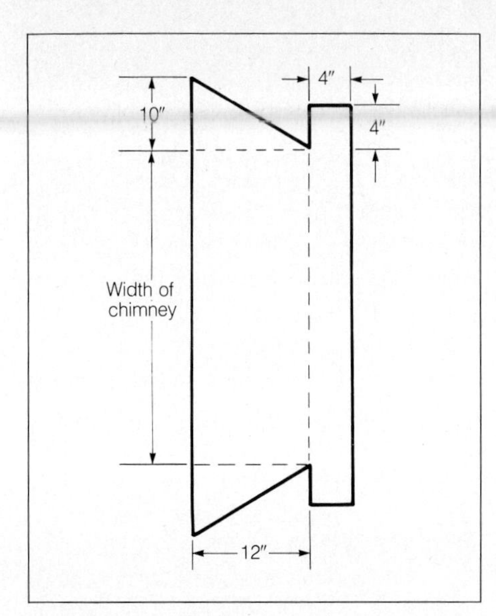

Flashing for the front face of a chimney must be cut to fit. This piece is cut to provide a 4-inch base over the roof shingles.

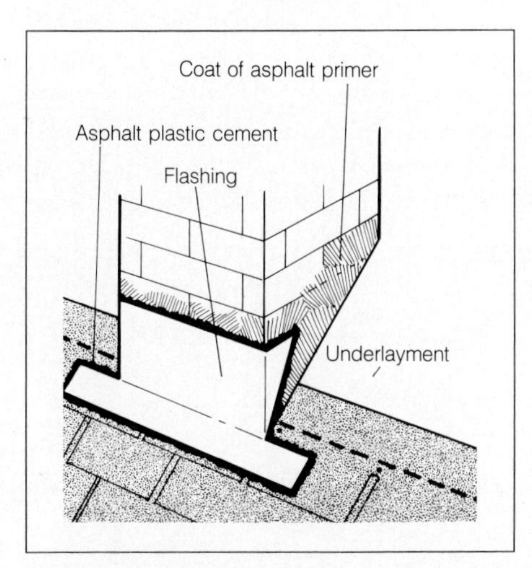

The flashing is bent so the 4-inch strip fits over the shingles and the diagonal sections fit on either side of the chimney.

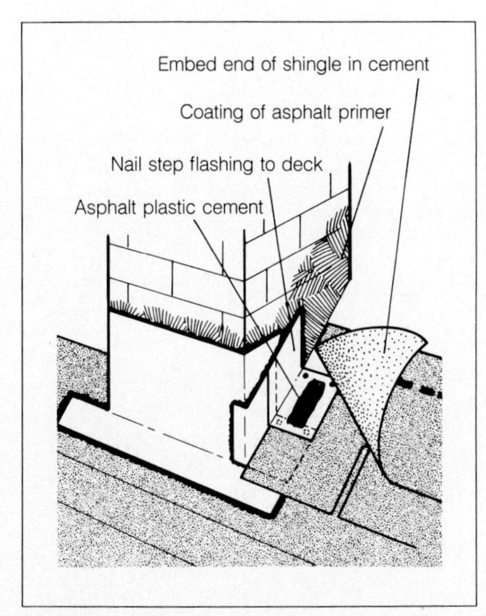

Step flashing is run up either side of the chimney. The flashing is nailed to the roof and covered with cemented down shingles.

Installing Base Flashing

Step one: fastening. Apply a coating of plastic asphalt cement where the top edge of the flashing meets the masonry. Drive roofing nails into the deck, through the lower edge of the flashing. Cover these nails with the cement. Cement the shingle ends down to the horizontal portion of the flashing.

Step two: the courses. Repeat the above procedures for each course. The flashing units should overlap each other by 3 inches. The upper flashing goes on top of the lower (preceding) one in each case.

Step three: finishing. When using this method instead of roll roofing base flashing, you also will need to apply metal cap flashing.

Cap or Counter Flashing

Counter flashing consists of 16-ounce or heavier sheet copper or 24 gauge galvanized steel with enamel finish, applied from the mortar joints of the brickwork to cover the base flashing so that water cannot enter from the top. Counter flashing comes in four sections. Although the flashing itself rarely deteriorates, crumbling mortar in the top joints can cause problems. This condition can be repaired as discussed below.

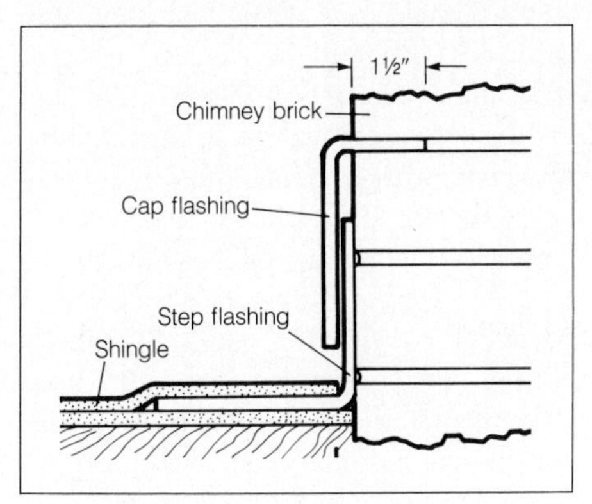

Cap flashing is installed with upper edge sealed in the chimney mortar. The lower section overlaps the base step flashing.

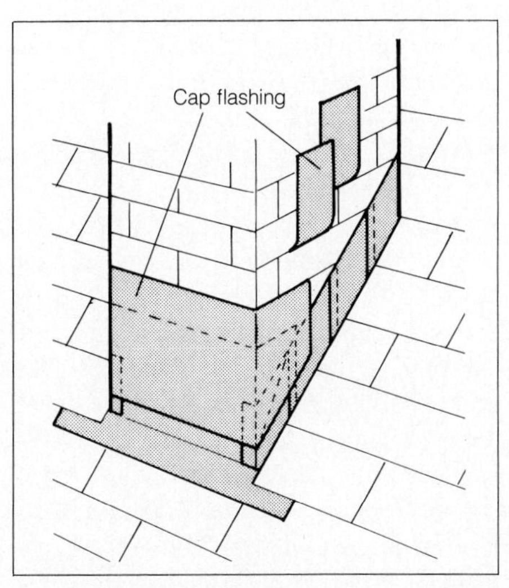

Cap flashing, which overlaps the base flashing by reaching from a mortar joint above, is shown lifted to indicate relative positions.

Step one: cutting the flashing. Use the old flashing, if roofing an existing house, as a pattern to cut an identical replacement. Or save the old flashing if it's in good shape.

Step two: front unit. The front section is one continuous piece, and it is installed first. Attach the bent top edge to the mortar joint and the lower edge to overlap the base flashing by 6 inches.

Step three: treating mortar joints. Rake the joint, using a small tuckpointing brick chisel, to a depth of 1½ inches. Bend 1½ inches of the top edge of the flashing and insert it into the cleaned-out area between the joints. Refill the joint with portland cement mortar or asphalt plastic cement.

Step four: side units. Cut several sections for each side in order to fit the locations of mortar joints and the roof pitch. Install the side units so they lap each other by at least 3 inches. Bend and embed their tops as discussed above. Secure with lead plugs.

Step five: back units. The back sections are cut similar to the side, and the back flashings at the corners bend around to overlap the chimney sides by 3 inches. Next come the two pieces of side flashing, embedded in the mortar in step fashion and secured by lead plugs.

Adding a Cricket

This diverts water and prevents accumulation of snow and ice. The inclusion of a cricket is important and is located behind the chimney.

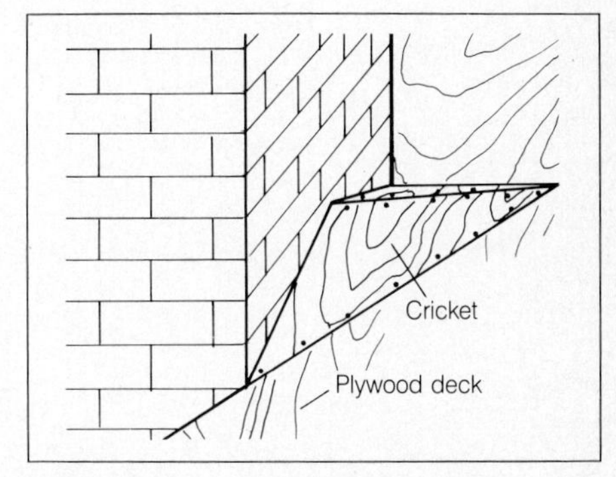

A cricket diverts rain and snow from the back of the chimney by creating a steep pitch that prevents backup and directs runoff.

Step one: building a cricket. A properly designed cricket is half as long as the width of your chimney. After determining the length of the cricket (front to back), cut a piece of 2x6 to that dimension and bevel one end so that it is flush with the roof slant when the cricket top is horizontal. Measure the distance between the bottom of the 2x6 and the point where the chimney back meets the roof. Cut a piece of 2x4 to that measurement and bevel the bottom end to fit the roof slant. Toenail both pieces to the roof with 8d nails. Nail the top piece to the bottom piece with 16d nails.

Use ¾ inch Exterior plywood to form the "roof" of the cricket. Cut two triangular pieces to extend from the frame to the outer edges of the chimney. Plane and/or saw the edges of the plywood to fit tightly at all joints. Cut out a piece of flashing to extend over the cricket four inches on all sides and up the chimney. If necessary, use two pieces of flashing and carefully solder the joints.

Step two: cricket installation. Place it upside down between the roof deck and back face of the chimney. Nail it down. Before you begin placement of the felt flashings, run shingles just to the chimney's front face.

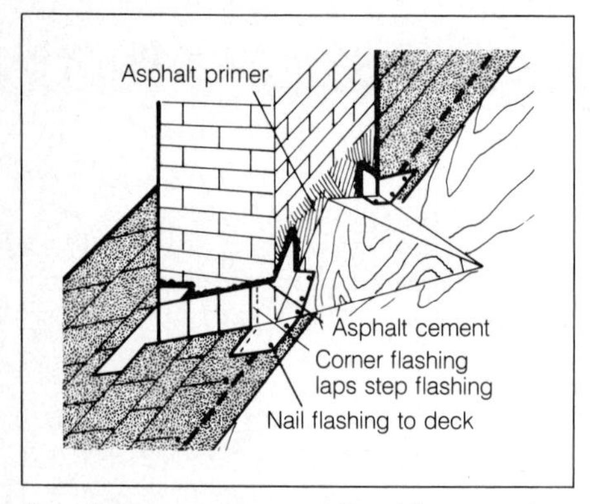

Base flashing runs around the chimney and up the cricket. Asphalt roofing cement seals the top (open) edge of the flashing.

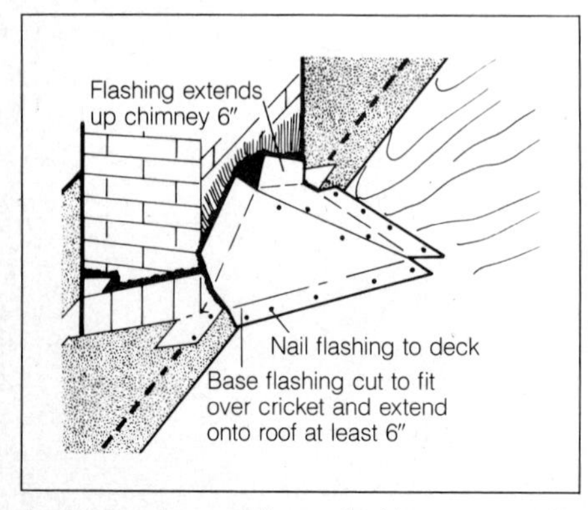

A single piece of base flashing covers the surface of the cricket and overlaps base flashing that comes up the chimney sides.

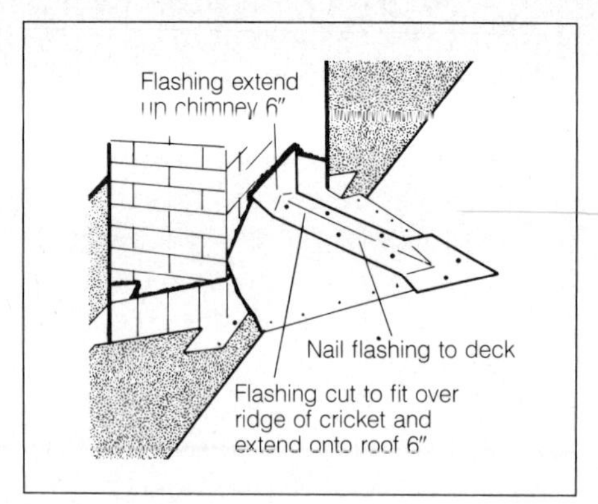

The peak of the cricket is double flashed with a separate strip that extends over the main roof surface to seal the joint.

Step three: sealing the surface. Apply a layer of asphalt primer to the brick to seal all surface areas. Plastic cement later will be applied.

Step four: installing the front of the base flashing. Cut out the base flashing as shown in the accompanying pattern. Apply a coat of asphalt plastic cement to the shingles around the chimney. Lay the lower section of the flashing on top of the shingles, in the bed of cement. Secure the upper vertical portion to the masonry by driving masonry nails into the chimney's mortar joints. Bend the triangular ends of this upper section to go around the chimney corners. Cement the ends in position.

Step five: installing the side base flashings. Following the pattern given, cut the side base flashings. Bend them to fit and install it so that it runs 10 inches up the chimney, with 8 inches on the deck and 4 inches in between. The base flashing sides lock into the front flashing by means of overlapping flanges. Fasten them to the deck using plastic cement and attach to the brick with nails and plastic cement. Cement the triangular ends of the upper section, which turn around the corners, over the front base flashing.

Step six: flashing over the cricket ridge. Cut and bend the base flashing so it will cover the cricket. Extend it anywhere from 6 to 12 inches up the brickwork. The deck section will project to cover some of the side base flashing. Cut another, rectangular roll roofing piece—with a "V" cut out from one side—to fit the cricket's rear angle. Set this piece in plastic cement. It should be centered over the cricket flashing that extends onto the deck. This adds protection where the deck meets the cricket ridge. Cut another, similarly shaped piece of flashing, with a V shaped cutout to fit the pitch of the cricket. Embed it in asphalt plastic cement, against the chimney and on top of the cricket ridge. Now apply a liberal coating of the plastic cement to all areas where the vertical sections of the base flashing meet the brick.

Step seven: shingling. If the cricket is a large one, cover it with shingles. If this is not suitable, place the end shingles in each course over the rear flashing and cement them in position.

Vent and Stack Pipes

All roofs will have at least one vent or stack pipe from the bathroom plumbing. There are ways to flash these with roll roofing and regular metal flashing, but most roofers use flashings made especially for this purpose.

Keep laying shingles until you reach the vent pipe. Take one shingle and cut a hole in it so it fits over the pipe. Set the shingle in a bed of roofing cement. Then

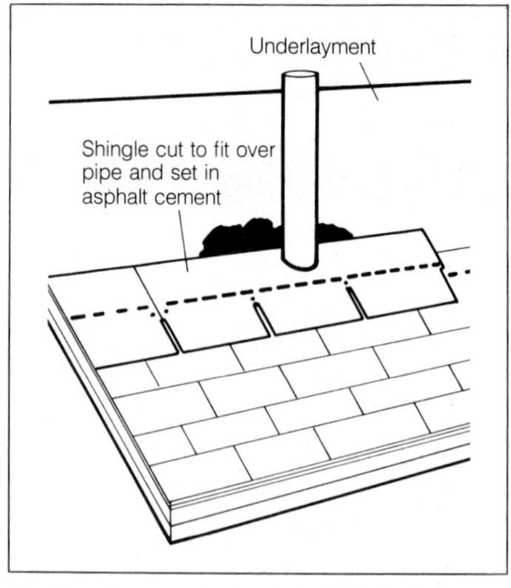

Flashing for a vent pipe is installed after a row of shingles, cut to fit around the pipe, is installed and sealed down.

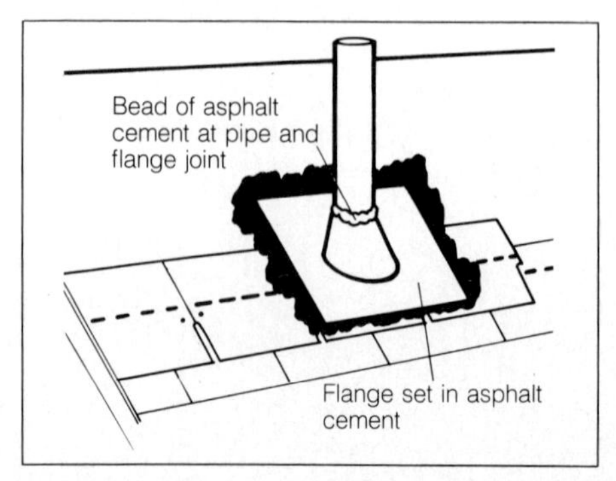

Conical flashing with base flange is fitted over the pipe and sealed down on the shingles. Joint between cone and pipe is sealed.

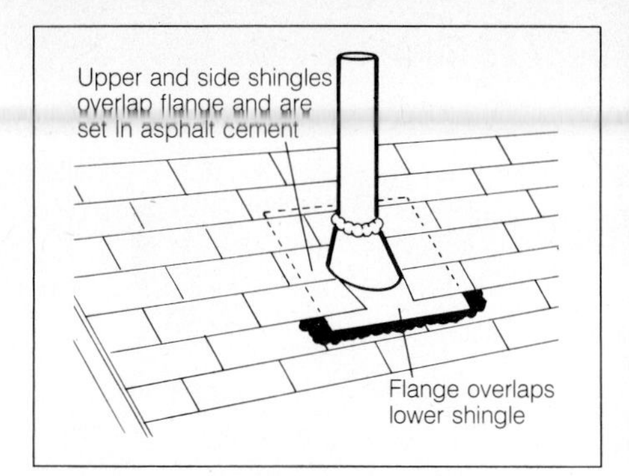

Upper courses of shingles are then laid over the flange and around the cone so water will flow smoothly down the roof.

place a preformed flashing flange over the shingle. It will fit snugly around the pipe. Again, set it in roofing cement. The flange should lie flat. Apply a bead of cement between the pipe and the flange.

As you lay the remaining shingle courses, cut the shingles so they fit around the pipe. Place each one in a bed of cement at the point where they meet and overlap the flange. Do not use too much cement; it can result in blistering. Fasten with nails, but do not place them too near the pipe. The lower edge of the flange will overlap the shingle course below; the upper edge of the flange will be covered by shingles.

At the ridge. If the vent or stack is at the ridge, the procedure is the same except that you lay the shingles from both sides until they come just up to the pipe. Bend the flange across and over the ridge so it lies on both roof planes. It will overlap the roof shingles at all points. Then position ridge shingles so they cover the flange. Where they overlap the flange, embed the ridge shingles in roofing cement.

This attic vent is flashed in the proper way. The upper edge is covered by the shingles; the lower edge overlaps the shingles.

Shingles are cut to fit around an exhaust vent pipe. Metal flashing will be added before the next row of shingles is laid so that upper shingles will overlap the flashing.

Flashings around Fans

Fans, skylights and similar obstructions usually come with built-in flashing, which should be left undisturbed by a reroofing job. If the flashing is missing or needs replacement, check with the manufacturer of the product. Sometimes the simplest way to fix such a problem is to improvise with aluminum tape or metal flashing similar to that used around chimneys. Be sure, however, that the bottom of the flashing rests on the shingles rather than under them.

FLASHING FOR WOOD ROOFS

Flashing for wood roofs is, for the most part, similar to that for asphalt shingles. In valleys, however, sheet metal—crimped in the center and at the edges—can substitute for roll roofing. Most valley flashing should extend 10 inches on each side of the valley, requiring 20-inch-wide metal. The only exception is for wood shingle roofs whose pitch is 12 inches per square foot (45 degrees). In this case, 14-inch-wide sheet metal can be used, extending seven inches on each side. Shakes always require 20-inch valleys, no matter what the pitch.

Application Guidelines

The valley metal should not run more than eight feet without an expansion joint, which should overlap at least six inches from above. Where the valley metal meets the roof edge, the metal should

Metal flashing is used for wood shingles. It is usually 20 inches wide, cut in 8 foot strips. Strips overlap 6 inches at joints.

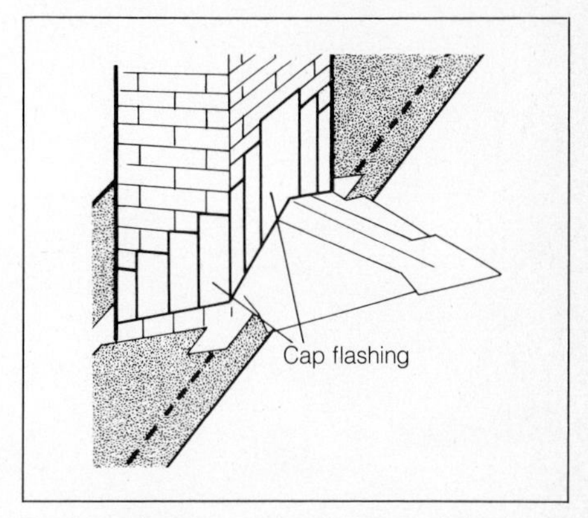

Final cap flashing at the chimney cricket is cut to fit over the cricket ridge, base and other cap flashing to divert runoff.

extend beyond the eaves out at least as far as the projection of the shingles.

Shingles or shakes extending into the valley should be sawed to the miter necessary to follow the line of the valley, butting on a perpendicular angle to the valley. Do not break joints into the valley, nor should you lay the shingles so that the wood grain is parallel with the centerline of the valley. Nail the shingles as far away from the valley center as possible.

Use wide shingles near the chimney flashing, and notch them at corners so that joint lines do not coincide with the sides of the chimney. Use copper or aluminum flashing, or galvanized steel with a factory-applied baked-on enamel.

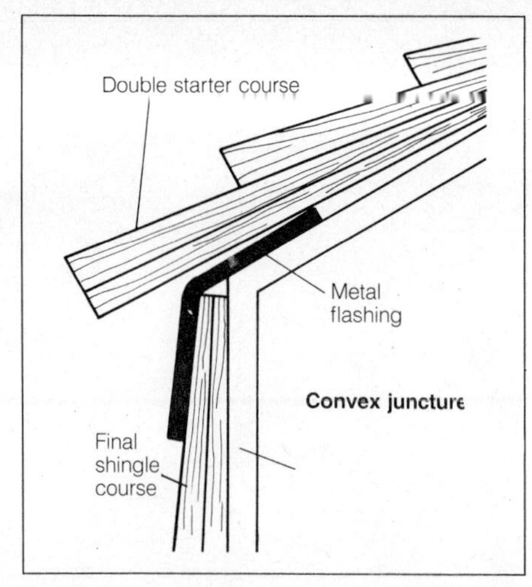

Angled metal flashing is used at the juncture of siding and roof. A doubled starter course overlays the flashing.

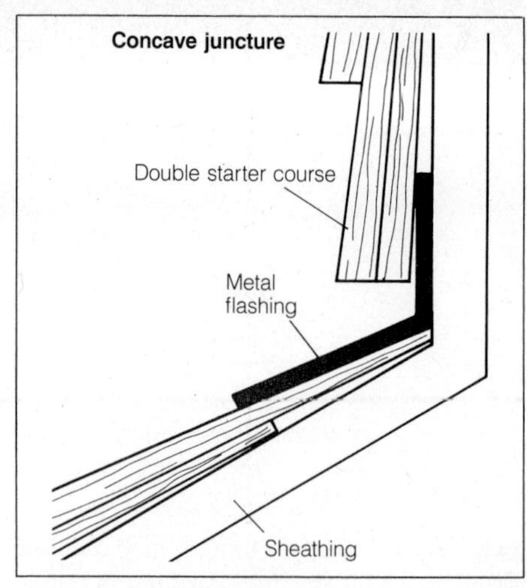

Flashing is also needed where a roof plane meets a wall. Double course siding overlaps flashing that overlaps roof shingle.

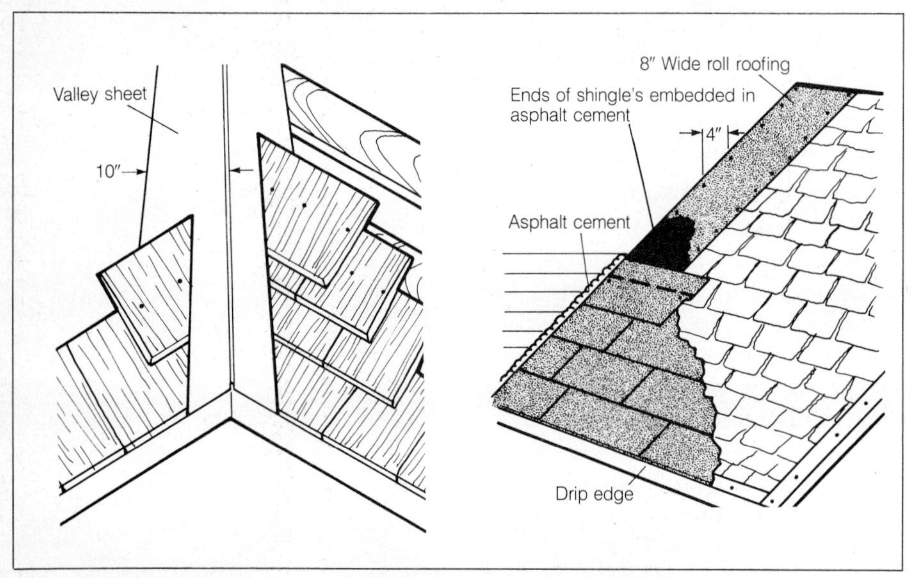

Roof plane junctures are handled in different ways depending upon the roofing material.

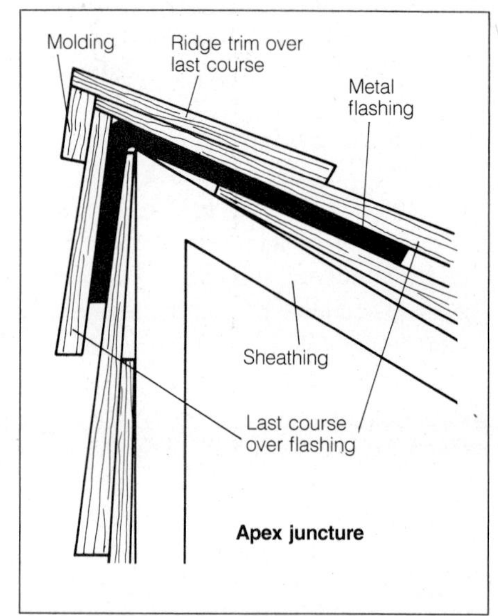

Use angled flashing at a peak.

When laying wood shingles over an asphalt roof, lay flashing first; then cover shingles with overlapping underlayment and lay the shingles.

VENTILATION

Condensation in the attic area is a fairly common problem, especially in cold regions during the winter. Warm moisture from the living spaces rises to the roof, meets the outside cold air; condensation forms. If severe enough, this moisture can cause rotting of the deck and framing. Wood roofs are not as susceptible, because the moisture can escape easier. Asphalt roofs are less permeable, and moisture can easily cause severe damage if not allowed to escape.

Insulation and Vapor Barriers

Fortunately, problems of this type can be prevented by adequate ventilation. Also helpful is adequate insulation with a proper vapor barrier. The insulation is used primarily to keep expensive heating and cooling in the living areas, where it belongs, although it does help some in keeping moisture out of the attic. More important is the vapor barrier attached to most insulation, which deflects moisture.

You should check your present ceiling insulation to see if it contains a vapor barrier. Look on the underside, next to the ceiling. If there is paper or foil there, you can presume that the vapor barrier is intact and doesn't need replacement. A few more words about vapor barriers. The vapor barrier always goes on the warm side of the wall, ceiling or floor. It is to be hoped that whoever built your house was smart enough to install the ceiling insulation with the vapor barrier down, but you never know. Floor insulation has the vapor barrier up, again on the warm side. Wrongly placed vapor barriers will actually trap moisture in the insulation and ruin the insulation.

Adding a vapor barrier. If there is nothing on the other side of the insulation, you are well advised to temporarily remove the insulation and install a vapor barrier beneath. (Do not add another layer of insulation, plus a vapor barrier, on top of the old insulation.) Sheets of thin polyethylene serve as good vapor barriers. Staple or nail it to the sides on tops of the joists.

Remember, when putting additional insulation over material that already has a vapor barrier, always buy "unfaced" insulation, without a vapor barrier. Two vapor barriers are worse than none at all. In no time, the insulation will become soggy with moisture and worse than useless. If your dealer doesn't have unfaced insulation, buy it with the vapor barrier, but either pull the barrier off or slash it heavily to allow moisture to pass through it.

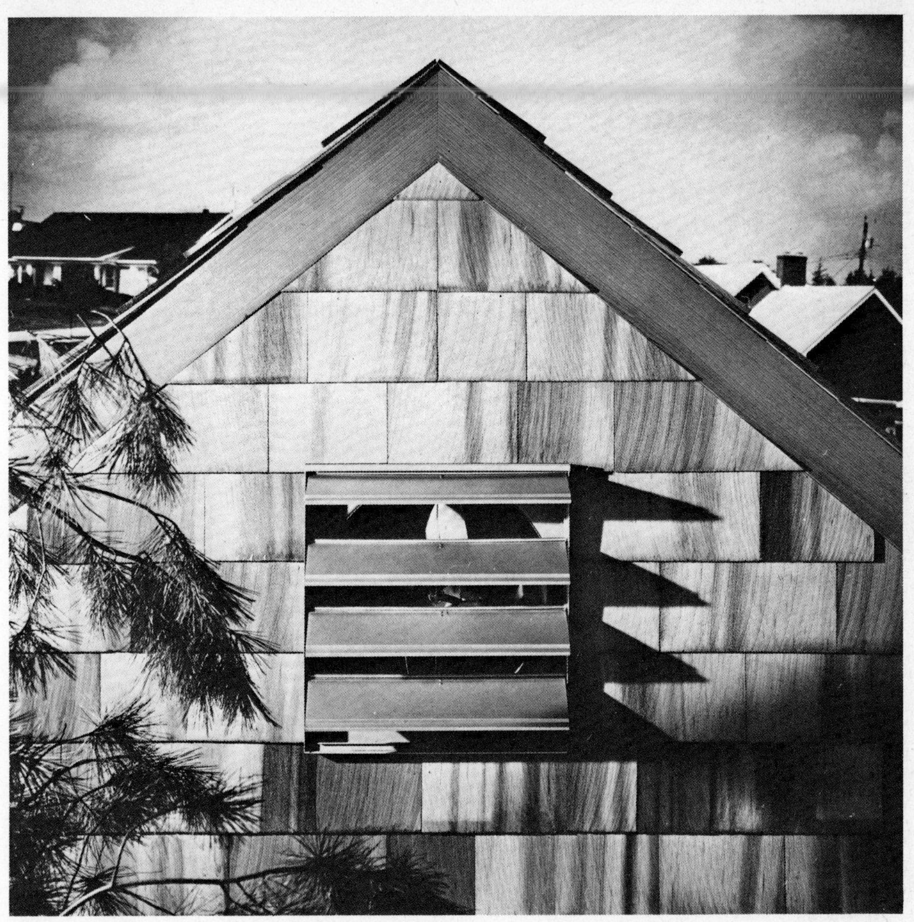

An attic fan with automatic opening louvers is a good investment for maintaining a comfortable home temperature during warm weather. Louvers close when the fan is not running.

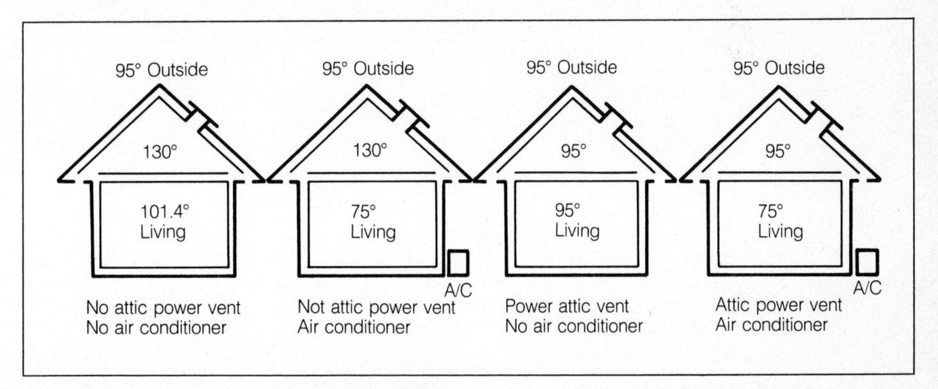

Air Changes Per Cubic Foot

The amount of ventilation needed within any space is based on "cubic feet per minute" or CFM. To translate this into CFM per square foot, keep in mind that for 10 air changes per hour a minimum CFM of 0.7 per square foot is needed. Tests have shown air temperatures in an attic can be reduced by 44.5 percent using 1.5 CFM per square foot, reduced by 67 percent when using 2.0 CFM per square foot. A ventilating rate of 2.0 CFM per square foot is the maximum suggested; higher rates do not contribute much to temperature reduction.

Where and How Many Vents

Vent openings must be sized and placed according to attic size and shape. The square footage ratio of vent to attic changes when grilles or screens or louvers are placed over the opening. It is best to use a "net free area" for the vent, comparing it to the attic floor area in square feet. The desired ratio is somewhere between 150 and 250 attic square

NET FREE AREA (SQ. IN.) TO VENTILATE ATTIC

Length in Feet	Width in Feet: 20	22	24	26	28	30	32	34	36	38	40	42	44	46	48	50
20	192	211	230	250	269	288	307	326	346	365	384	403	422	441	461	480
22	211	232	253	275	296	317	338	359	380	401	422	444	465	485	506	528
24	230	253	276	300	323	346	369	392	415	438	461	484	507	530	553	576
26	250	275	300	324	349	374	399	424	449	474	499	524	549	574	599	624
28	269	296	323	349	376	403	430	457	484	511	538	564	591	618	645	662
30	288	317	346	374	403	432	461	490	518	547	576	605	634	662	691	720
32	307	338	369	399	430	461	492	522	553	584	614	645	675	706	737	768
34	326	359	392	424	457	490	522	555	588	620	653	685	717	750	782	815
36	346	380	415	449	484	518	553	588	622	657	691	726	760	795	829	864
38	365	401	438	474	511	547	584	620	657	693	730	766	803	839	876	912
40	384	422	461	499	538	576	614	653	691	730	768	806	845	883	922	960
42	403	444	484	524	564	605	645	685	726	766	806	847	887	927	968	1008
44	422	465	507	549	591	634	676	718	760	803	845	887	929	971	1013	1056
46	442	486	530	574	618	662	707	751	795	839	883	927	972	1016	1060	1104
48	461	507	553	599	645	691	737	783	829	876	922	968	1014	1060	1106	1152
50	480	528	576	624	672	720	768	816	864	912	960	1008	1056	1104	1152	1200
52	499	549	599	649	699	749	799	848	898	948	998	1048	1098	1148	1198	1248
54	518	570	622	674	726	778	830	881	933	985	1037	1089	1141	1192	1244	1296
56	538	591	645	699	753	807	860	914	967	1021	1075	1130	1184	1237	1291	1345
58	557	612	668	724	780	835	891	946	1002	1058	1113	1170	1226	1282	1337	1392
60	576	634	691	749	807	864	922	979	1037	1094	1152	1210	1267	1324	1382	1440
62	595	655	714	774	834	893	953	1012	1071	1131	1190	1250	1309	1369	1428	1488
64	614	676	737	799	861	922	983	1045	1106	1168	1229	1291	1352	1413	1475	1536
66	634	697	760	824	888	950	1014	1077	1140	1204	1268	1331	1394	1458	1522	1585
68	653	718	783	849	914	979	1045	1110	1175	1240	1306	1371	1436	1501	1567	1632
70	672	739	806	874	941	1008	1075	1142	1210	1276	1344	1411	1478	1545	1613	1680

feet per square foot of venting. The more attic square feet per square foot of net free vent, the longer the complete air changes will take.

Now that you have some idea of the number of vents needed, you must consider placement of the vents, which is essential for efficient air replacement.

Wind pressure is one important factor. Wind striking a side of the roof produces a vacuum in that area which, in reaction, pulls the air back toward the house in an adjacent area. You need equal numbers of vents in both areas of vacuum (called negative pressure) and pulled-in pressure (called positive pressure). Vents placed in the vacuum area allow air to be drawn from the attic; vents in other, positive-pressure areas, pull air in.

Another influence results from the "hot-air-rises" principle. High vents let out escaping, heated air. Low vents replace hot air with new, cooler air from the outside.

Types of Vents

Louvers. In addition to vapor barriers, your attic should contain louvers to provide passive ventilation of air from the attic space. Louvers not only let moisture escape, they also provide an escape for the very hot air that builds up in attics during the summer.

Louvers are small domes that mount on the roof, near the ridges. They are available with screens to prevent insects from coming in. However, screens reduce airflow and can become clogged. Slit design screens are best because they avoid these problems.

Louvers in gable ends. In gable roofs, there should be louvers at the top of both end walls. These are triangular, or sometimes square. If this is the only form of ventilation, there should be one square foot of opening for every 300 square feet of attic floor area. Larger or additional louvers can be obtained from most building supply dealers.

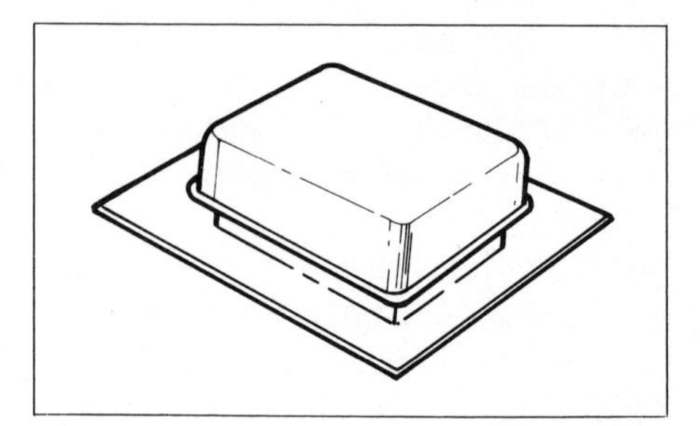

Static ventilators are equipped with complete flashing and have a housing designed to prevent rain or snow from entering the house.

Louvers are installed simply by cutting away the siding and sheathing between

POWER VENTILATOR REQUIREMENTS

Length in Feet	Width in Feet															
	20	22	24	26	28	30	32	34	36	38	40	42	44	46	48	50
20	280	308	336	364	392	420	448	476	504	532	560	588	616	644	672	700
22	308	339	370	400	431	462	493	524	554	585	616	647	678	708	739	770
24	336	370	403	437	470	504	538	571	605	638	672	706	739	773	806	840
26	364	400	437	473	510	546	582	619	655	692	728	764	801	837	874	910
28	392	431	470	510	549	588	627	666	706	745	784	823	862	902	941	980
30	420	462	504	546	588	630	672	714	756	798	840	882	924	966	1008	1050
32	448	493	538	582	627	672	717	761	806	851	896	941	986	1030	1075	1120
34	476	524	571	619	666	714	762	809	857	904	952	1000	1047	1095	1142	1190
36	504	554	604	655	706	756	806	857	907	958	1008	1058	1109	1159	1210	1260
38	532	585	638	692	745	798	851	904	958	1011	1064	1117	1170	1224	1277	1330
40	560	616	672	728	784	840	896	952	1008	1064	1120	1176	1232	1288	1344	1400
42	588	647	706	764	823	882	941	1000	1058	1117	1176	1234	1294	1352	1411	1470
44	616	678	739	801	862	924	986	1047	1109	1170	1232	1294	1355	1417	1478	1540
46	644	708	773	837	902	966	1030	1095	1159	1224	1288	1352	1417	1481	1546	1610
48	672	739	806	874	941	1008	1075	1142	1210	1277	1344	1411	1478	1546	1613	1680
50	700	770	840	910	980	1050	1120	1190	1260	1330	1400	1470	1540	1610	1680	1750
52	728	801	874	946	1019	1092	1165	1238	1310	1383	1456	1529	1602	1674	1747	1820
54	756	832	907	983	1058	1134	1210	1285	1361	1436	1512	1588	1663	1739	1814	1890
56	784	862	941	1019	1098	1176	1254	1333	1411	1490	1568	1646	1725	1803	1882	1960
58	812	893	974	1056	1137	1218	1299	1380	1462	1543	1624	1705	1786	1868	1949	2030
60	840	924	1008	1092	1176	1260	1344	1428	1512	1596	1680	1764	1848	1932	2016	2100
62	868	955	1042	1128	1215	1302	1389	1476	1562	1649	1736	1823	1910	1996	2083	2170
64	896	986	1075	1165	1254	1344	1434	1523	1613	1702	1792	1882	1971	2061	2150	2240
66	924	1016	1108	1201	1294	1386	1478	1571	1663	1756	1848	1940	2033	2125	2218	2310
68	952	1047	1142	1238	1333	1428	1523	1618	1714	1809	1904	1999	2094	2190	2285	2380
70	980	1078	1176	1274	1372	1470	1568	1666	1764	1862	1960	2058	2156	2254	2352	2450

the two studs and nailing them to the studs on each side. Flashing usually comes with the louvers, or can be supplied as explained above.

Louvers in the eaves. If there is room in the eaves, more ventilation can be obtained by adding small round louvers. These provide air movement because of differing temperatures, regardless of any wind or breeze outside. When soffit ventilation is provided, the total ventilation area can be reduced to one square foot for every 900 square feet of attic area. All vents and louvers should be screened to keep out bugs, but since screens cut down considerably on the actual venting ability, you may wish to double the relative footage of the eave vents so that you have two square feet instead of one for every 900 square feet of attic space. Screens should be as coarse as conditions permit—not smaller than #16. Lint and dirt will clog finer mesh. Never paint the vent screens.

Soffit vents. Hip roofs should always contain soffit vents. A continuous slot at least ¾ inch wide is the most efficient type of ventilation for a hip roof. In addition, there should be at least one large louver vent or several smaller vents near the ridge, preferably in the rear for aesthetic reasons. Because this vent is the only one to provide air circulation near the floor, it is effective and helpful in the face of all types of wind conditions and directions. However, the roof area is not cooled if only these vents are used.

How to Install an Attic Fan

The cooling power from an attic fan is more than you might think, and an attic fan can lower air conditioning bills considerably during the summer months when the night air is mild (not hot). An attic fan pulls outside air in through open windows and exhausts the hot air in the attic. This air then filters out through attic gable and soffit vents. The fan circulates the air, and the air movement cools you via evaporation of perspiration from your skin.

Attic fans can move lots of air fast. Each mode has a selection of capacities of that they can handle. Figure the amount of floor space in your home and buy a fan that can handle it.

The fans come in two basic sizes. One unit is made for installation in the gable end of the house. The other model sits between ceiling and joists. You must cut a hole in the ceiling, house siding, or roof and cover it with screening or louvers for the fan to work efficiently.

The ceiling-mounted fan is probably

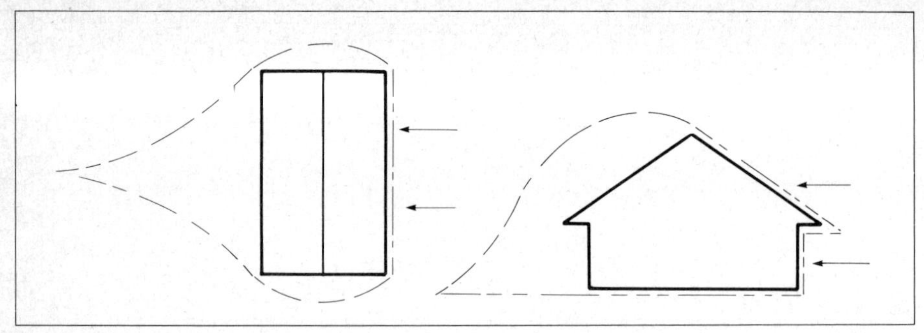

When air flows over any object, there is a change in the air speed and air pressure. The speed increases and the pressure decreases as the air passes over the object.

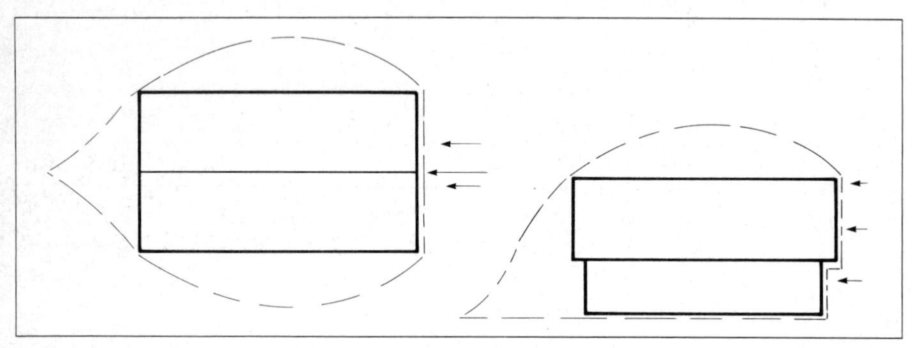

Air hitting a gable end blows through a louver; pressure change pulls it out the other end.

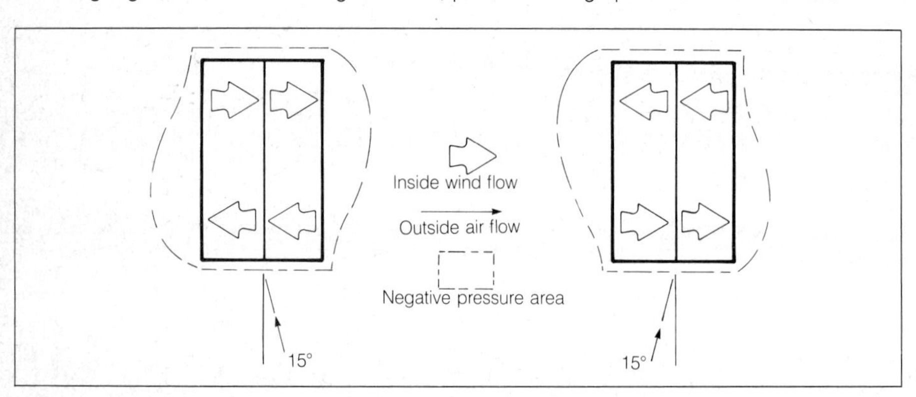

The negative pressure created by the wind flow helps pull air through static ventilators and increases the effectiveness of power-driven ventilator units in louvers or roof fans.

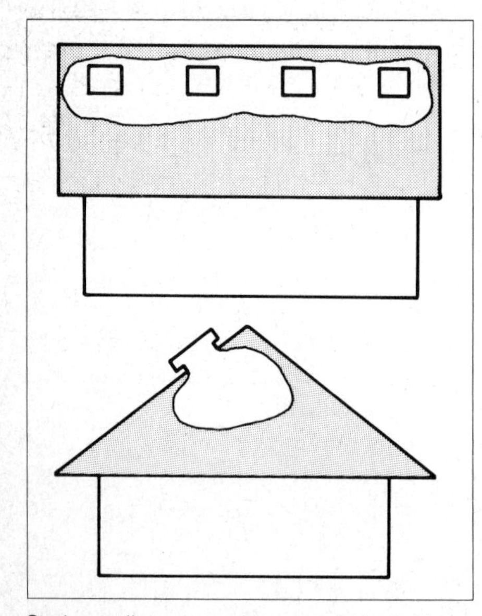

Static ventilators must be spaced along a large roof expanse for good ventilation because they draw air from a limited area.

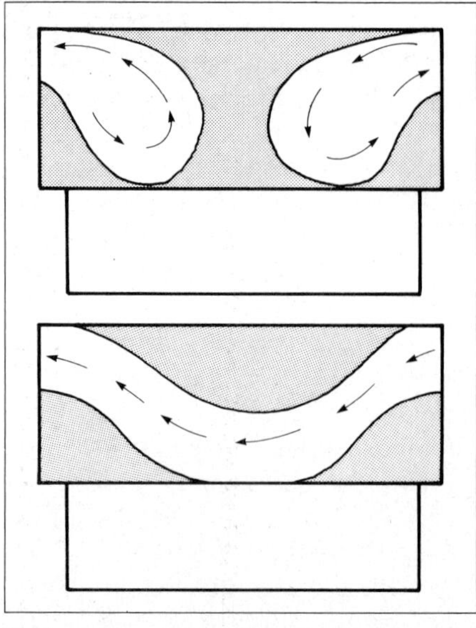

Gable-end louvers circulate air in and out of the same end unless wind is moving directly in line with the length of the house.

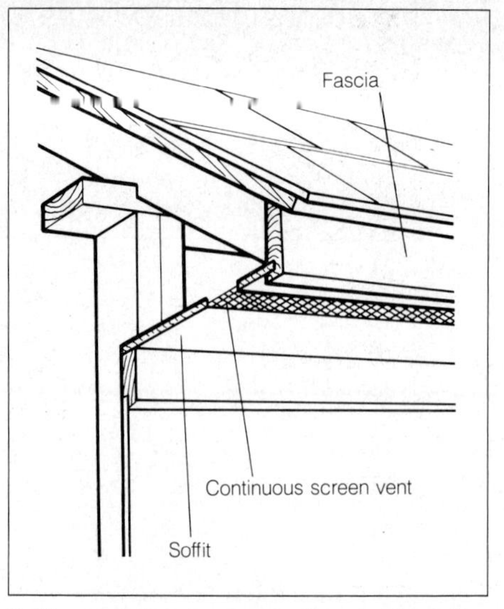

Soffit ventilation increases the air circulation through the louvers. One common means of soffit ventilation uses a continuous screen.

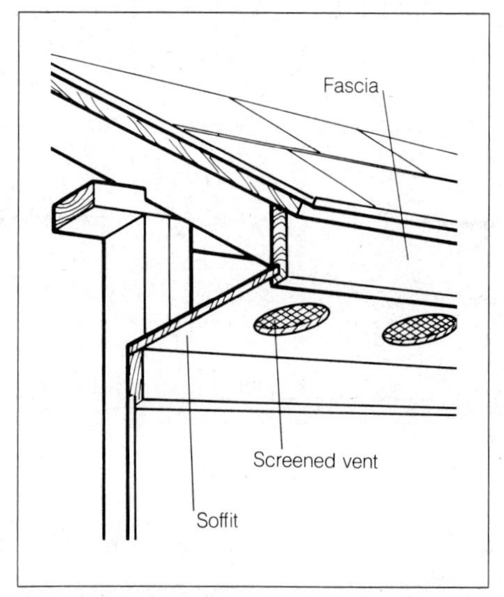

Soffit ventilation may be added by installing a series of screen-covered holes. The holes allow air to flow through the soffit.

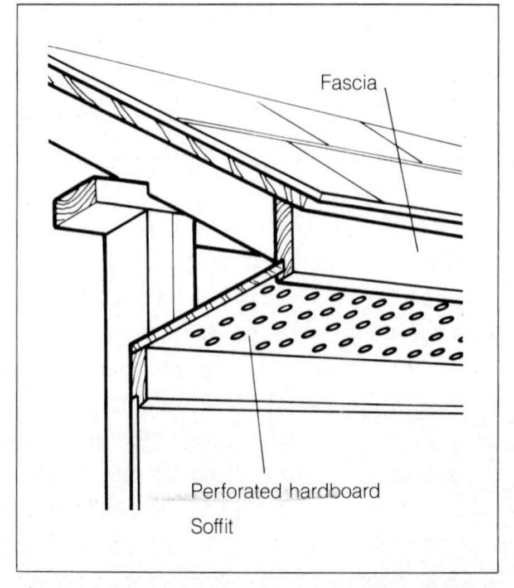

Another method of providing soffit ventilation substitutes a perforated hardboard panel for the usual soffit board.

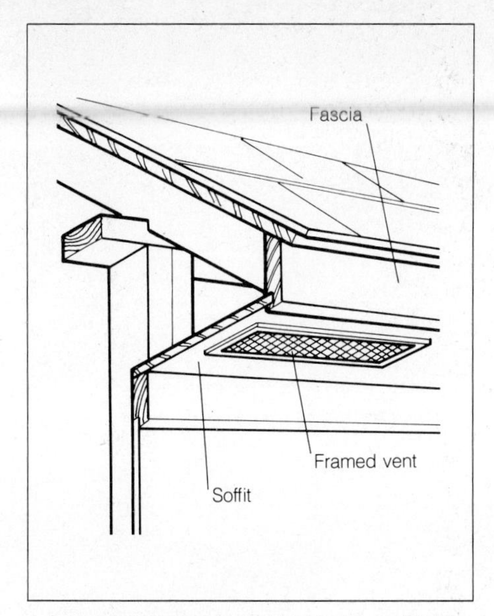

A framed vent may be added to the soffit. Use siding vents under soffit, but above attic floor, for attic connection and cooling.

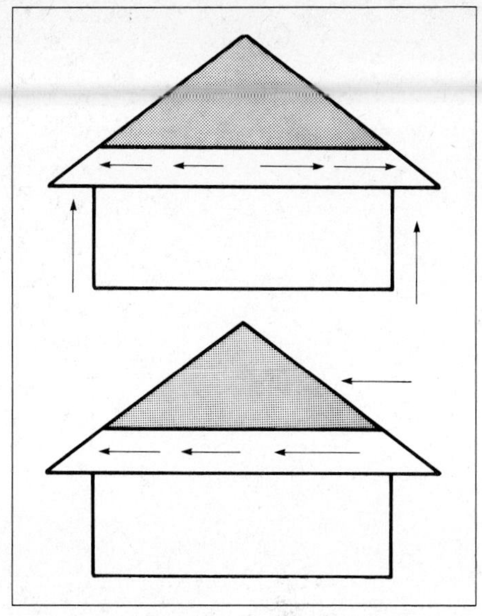

Without attic connection, soffit vents allow continuous circulation of air at the soffit but do not aid in attic ventilation.

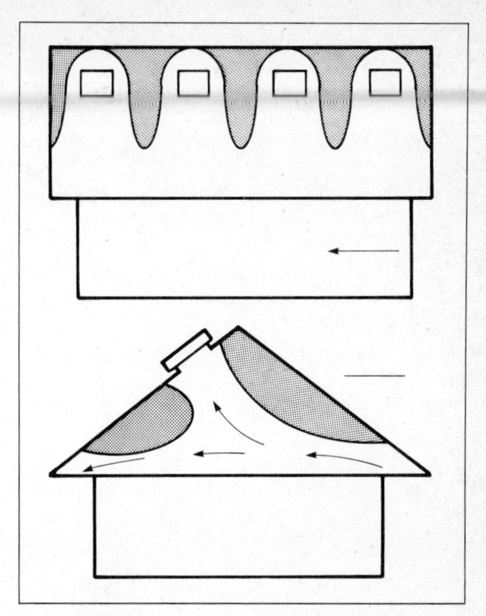

Soffit and roof vents will work well together, but the air flow will still tend to leave areas of unexpelled air.

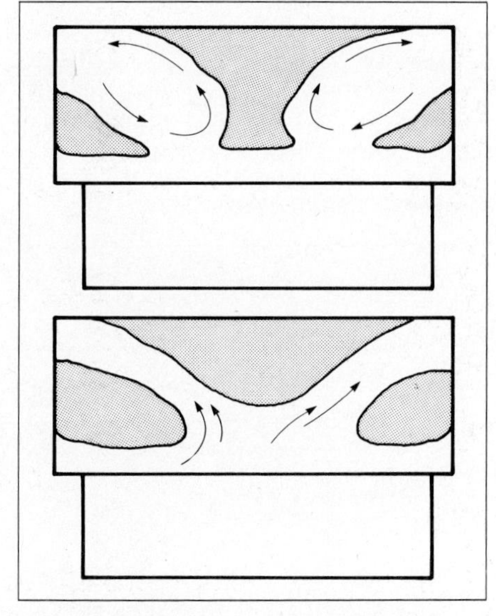

Addition of soffit vents increases the efficiency of the gable louvers if a passage to attic is provided for air exchange.

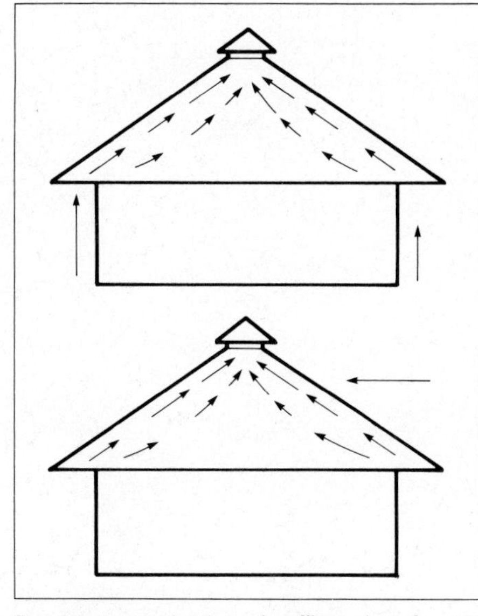

The best combination of soffit and roof vents puts the roof vent near the ridge or peak of the attic, where air naturally rises.

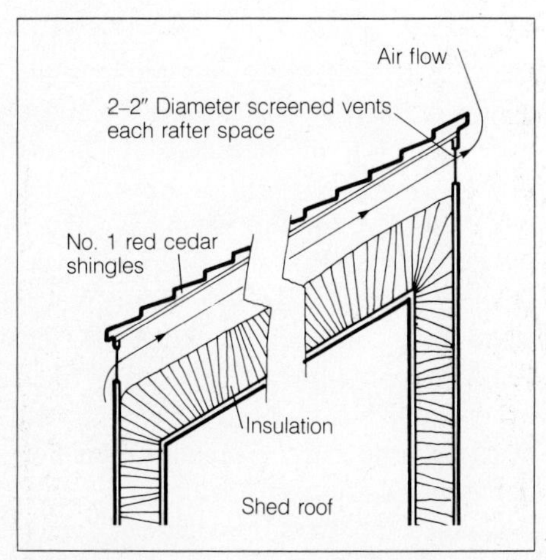

Effectiveness of insulation is increased when upper areas are well ventilated in order to prevent excess heat buildup.

easier to install, although the wall-and-roof model is not difficult. Manufacturers of both types of fans offer installation instruction. You will find guidelines here, although you should follow instructions packaged with the fan.

Tools and materials. You will need a hammer, square, tape measure, 2x8 or 2x10 headers (one piece 4 feet long), screwdriver, 16d common nails, armored cable and wire, junction box, razor knife, pliers, power jigsaw and crosscut blades or keyhole saw, brace and ½ inch bit.

Placement. Locate a spot for the fan in the ceiling that separates the attic and the room below. Hallways are the most popular because they usually are central to the rest of the house and pull air through all the rooms. You can control air flow by opening and closing windows and doors.

In the attic or crawl space, measure the opening between the ceiling joists and mark this opening as a guide for the saw. The opening's size will depend on the size of the fan.

Cutting the opening. Make the cut in the attic floor (and room ceiling) with a power jigsaw—or use a keyhole saw to bore a hole in the ceiling to assept the saw blade. Clean up the mess from the gypsum wallboard on the floor below immediately. Check the louver panel against the hole to see if it fits properly. If not, trim the ceiling material with a saw or razor knife to obtain the proper fit.

Supporting the fan. Tie off the ceiling joists with headers that are the same size as the ceiling joists, usually 2x8s or 2x10s. Square the headers in the space and spike them to the joists.

Installing the fan. Position the fan between the joints and headers. It should slip easily into the space. You may have to build a square or rectangle of 1x3 or 1x4 strips on which the fan can rest. This depends on the model; consult the directions.

Assemble the fan unit; place it over the ceiling cutout and fasten the fan in position using the nuts and bolts and screws furnished with the fan.

Go into the room below and install the louver vent or grill in the ceiling. It probably fastens to the fan housing. If not, you can screw it to the 1x3 or 1x4

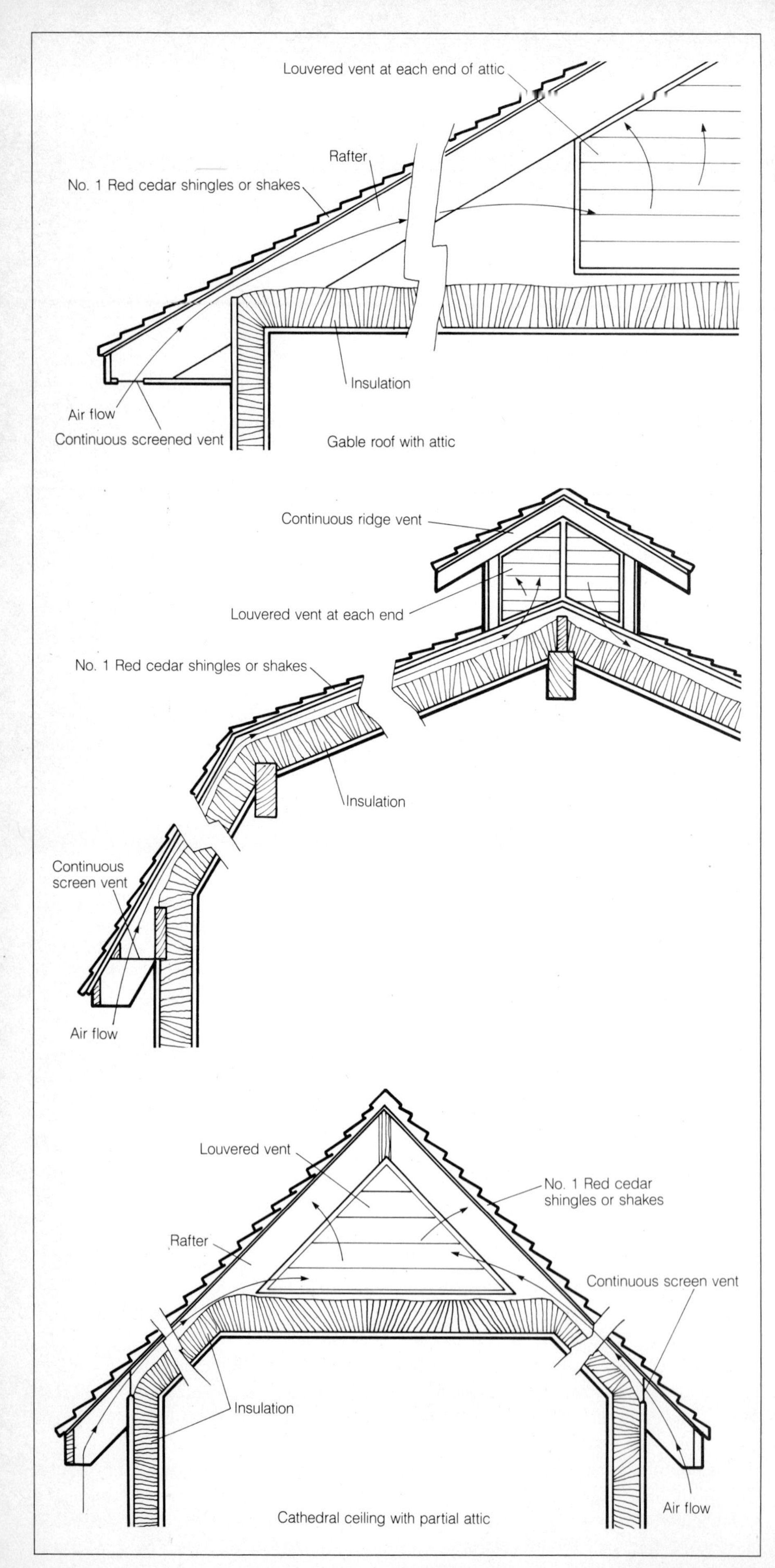

Power ventilators come in many styles and designs and can be installed as easily as static ventilators. Flashing is included.

Cut an opening between rafters. Seal upper flashing under shingles and lower flashing over shingles to install power ventilator.

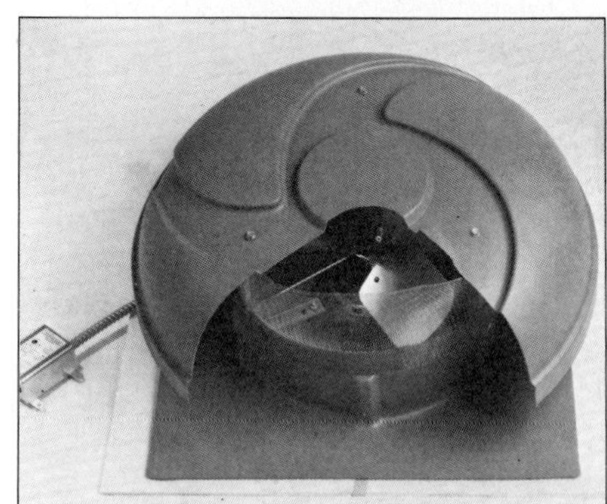

Power ventilators have a fan and motor, flashing, a screen to keep out insects, and a protective housing to keep out rain or snow.

frame that you made, up through the ceiling at the edge of the cut-out. Or tack it to the joists and headers through the ceiling.

Connecting the fan wiring. To finish the job, connect the fan to electric power. Before you start work, turn off the power at the main fuse box or circuit breaker. Run BX cable from a nearby wall switch to a junction box fastened to a ceiling joist near the attic fan. A prewired junction box is usually furnished with the fan unit.

Connect the wires from the wall switch to the wires in the junction box, using wire nuts, or according to the wiring diagram and information supplied with the fan.

10

GUTTERS AND DOWNSPOUTS

Gutters catch rain flowing off the roof and carry it to the downspouts (leaders), which disperse it harmlessly away from the house through drain pipes or natural grading. Pools of water that collect near the foundation, streaked house paint, damp basements or ridges in the ground under the eaves—all these are signs that the gutters are not working properly. Many times, inspection will reveal a ball or toy blocking the downspout, or leaves that have accumulated in the gutters and drains. Removal of the foreign material ordinarily will cure the problem. If left unattended over a longer period, clogged leaders or gutters can cause a torrent of rain in one or two areas. Soil erosion and other problems can result. Rainwater or melted snow can back up into the house.

MAINTENANCE AND PREVENTION

Inspections and maintenance each spring and fall can prevent any serious gutter problems. In addition to removing leaves and other debris, downspouts should be cleaned, using a plumber's snake if necessary.

Possible Solutions and Repairs

Strainers. Strainers designed for the top of downspouts should be installed. They are available in most hardware and building supply stores. By keeping gutters and leaders clean, backups and consequent rusting should be prevented. Where there are a lot of trees above, strainers should be placed over the gutters as well as on the leaders. Mesh sold for this purpose hooks onto the outside of the gutters and underneath the first row of shingles. Before removing old strainers in gutters or leaders, clean away surrounding debris so it does not fall into the downspouts.

Surface care. Gutter maintenance includes painting them with special gutter paint as necessary, and coating the insides of the gutters with roofing cement every few years as needed. Small holes and cracks should be covered with roofing cement or silicone caulk. Wire-brush loose metal and corrosion from rusted gutters, then wipe clean with a cloth. Spread roofing cement over the rusted area.

Straps and hanger. Check for loose straps or gutter nails. Check for correct drainage by pouring a bucket of water into the ends of each gutter. If the water does not flow smoothly downhill to the leaders, adjust the hangers or nails to provide the proper slope. Hangers can be renailed higher or lower on the roof using

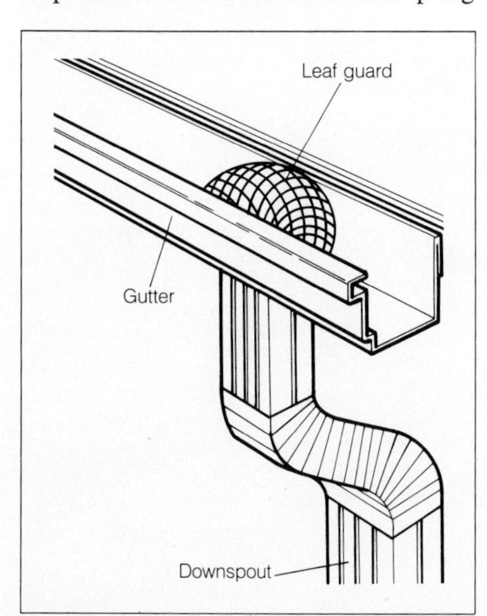

To prevent accumulation of debris, especially leaves, add a ball-shaped leaf guard to keep downspouts clean.

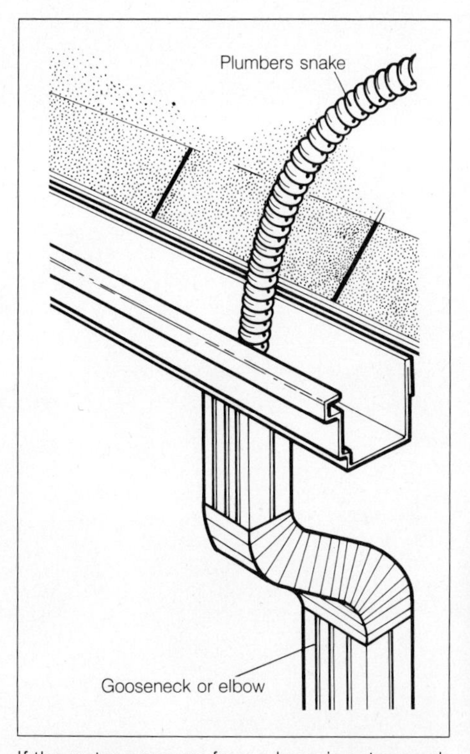

If the water pressure from a hose is not enough to dislodge collected leaves or other debris, use a plumber's snake.

Before repainting an old gutter, scrape off the old layers. Do not paint new metal gutters without first priming the surface.

new galvanized nails. If spikes are used through the gutter and cannot be otherwise removed, cut through the nail at the end of the ferrule and attach with new gutter spikes, available at most hardware dealers. Sometimes the nails or hangers can be bent to the shape needed.

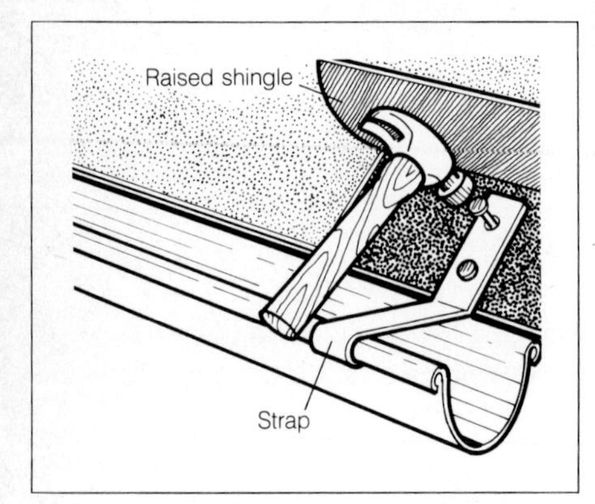

Check all your gutter systems for loose straps. Adjust them and renail when necessary.

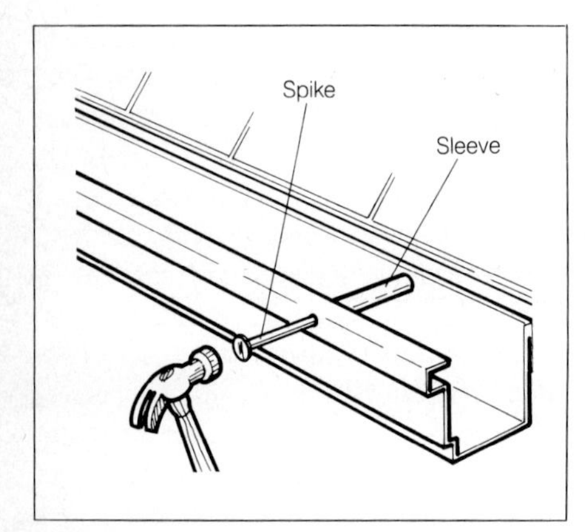

Renail gutter spikes, adjusting them as needed. If the spike cannot be removed, cut it off with a hacksaw and drive in a new one.

Leaks. For larger leaks, look for a gutter-patching kit at your dealer, or cut out a piece of fiberglass, canvas or thin aluminum to serve as a patch. Coat the area around the hole with roofing cement and lay the patch over this. Press it down firmly into the cement, then apply another coat of roofing cement extending an inch or two beyond the patch. If the leak is quite small, you can just press a piece of adhesive-backed aluminum tape over the hole.

Splash blocks. Another problem is created when the downspout doesn't empty into a storm sewer or other drainage system. The accumulated rainwater can form pools around shrubbery or even cause leaks into the basement. A concrete splash block, available at masonry and building supply houses, will cure most problems of this type. Sometimes, how-

To repitch gutters so water drains properly, bend the hangers with pliers. If the hangers are spikes/ferrules, pull the spikes, adjust the gutter and replace spikes.

For cleaning debris from downspouts, use a garden hose as a water jet. Install gutter guards and downspout strainers to keep leaves out of the water-carrying system.

Caulk the joints of gutter end caps to prevent leakage. Caulk joints of slip connectors on the inside of the gutters. Use butyl caulking for this repair.

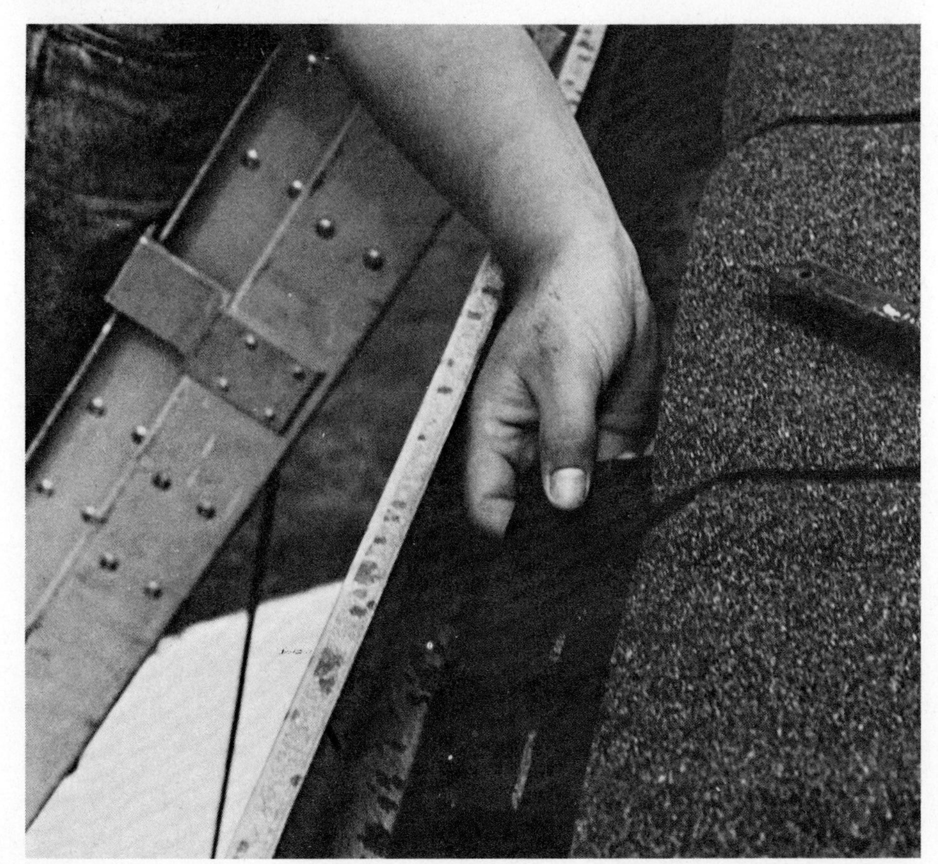

To temporarily patch a gutter, clean the gutter and spread asphalt roofing cement over the hole. Embed a piece of asphalt building paper or roll roofing over the hole.

ever, regrading or installation of a drainage system may be necessary.

NEW GUTTERS AND DOWNSPOUTS

Gutters eventually fall into disrepair. The fascia, or other wood to which gutters attach, often decays first. This results in loosened gutters and can bend them out of shape, or cause them to fall from the roof.

If one hanger or spike comes loose, it should be repaired or replaced promptly, since the weakness in that part of the gutter will eventually place too great a strain on the entire system, and the whole gutter will eventually bend and/or fall. It is almost impossible to repair a gutter once it has been bent out of shape.

Replacing a Section

If one gutter section has been bent beyond salvation, and the rest of the system is still in good shape, you can usually replace just that section. When corrosion is the culprit, and all gutters appear to be the same age, you may as well replace the entire system, since the other sections are certain to be vulnerable to rust and decay in the near future.

Remove the gutter(s) by taking off the hangers or spikes. Make sure that no one is below, or that nothing valuable is underneath, because the loosened gutter will be difficult to hold and may fall to the ground. To prevent this, get a helper and another ladder.

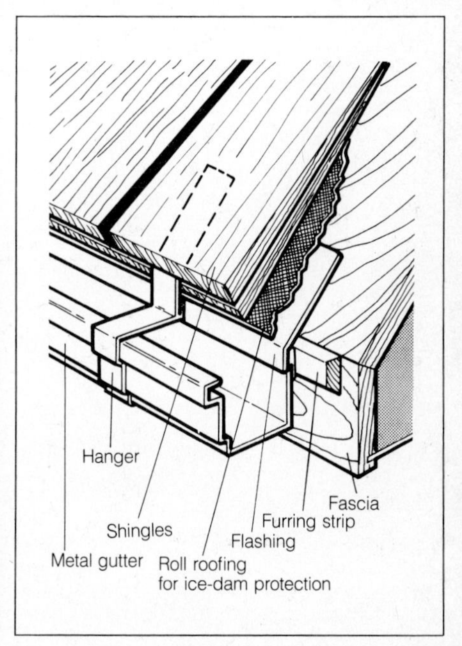

One end of the gutter lies under the drip edge of the shingles; the hanger slips between shingles of the doubled starter course.

When the gutters have all been removed, check the condition of the wood to which the new gutters will be attached. Replace any wood which shows signs of decay.

Take a piece of the old gutter to a dealer to help you find the same kind for replacement. Gutter sections attach by means of slip joints, and one type will not fit into another. Note whether the section to be repaired is in the center, at the end, or empties into the downspout. Gable roofs ordinarily have two separate systems for each side of the house, but hip and some other roofs sometimes have a continuous system. This will involve corner or perhaps other sections, which may be difficult to find at a small dealer. After slipping the new section into place, coat the joint with roofing cement to prevent leaks.

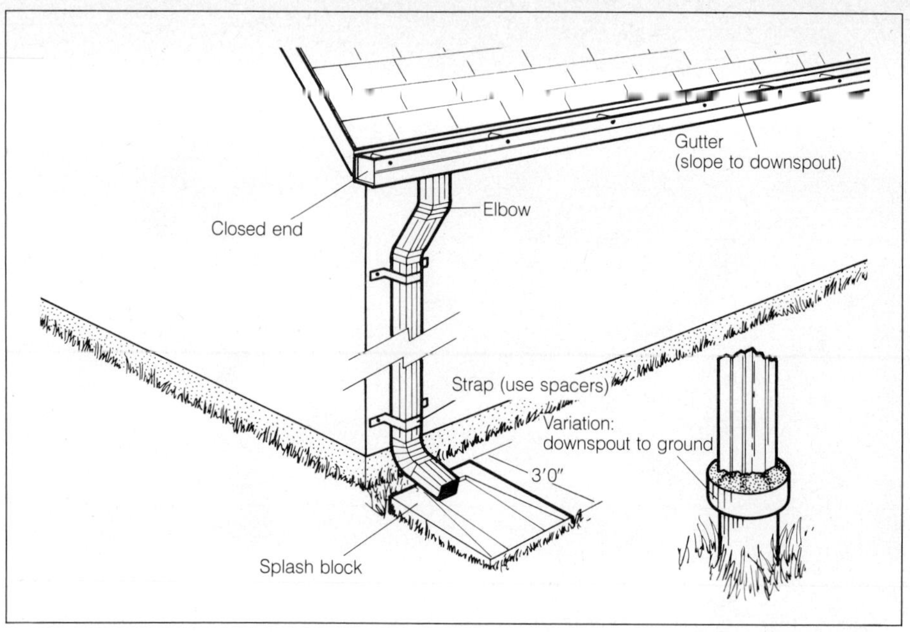

Shown at left is a typical gutter installation and connection to downspouts. Shown at right is how to attach the downspout to the drainage system.

Replacing Gutter System

When the whole system needs replacement, you have a choice of types of gutters and materials. You also can improve the system if needed by putting in wider gutters and downspouts or adding more leaders if indicated. Two downspouts are usually sufficient on gable roofs with two separate gutters, but hip roofs with a continuous gutter should have four leaders.

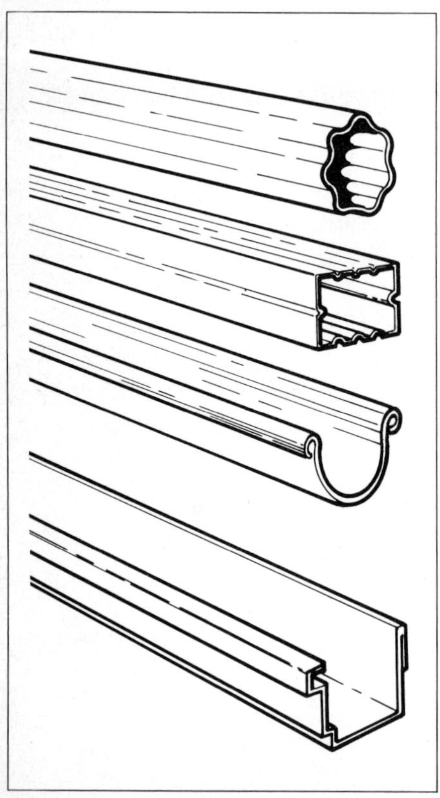

From top to bottom are gutter and downspout types: round downspout; rectangular downspout; half-round gutter; "OC" formed gutter.

Choosing new gutters and downspouts. The size of the gutters depends on the size and spacing of the downspouts. One square inch of downspout is required for each 100 square feet of roof. When downspouts are spaced up to 40 feet apart, the gutter should have the same width as the downspout. For greater spacing, the width of the gutter should be increased.

Most gutters are made of aluminum, but copper and galvanized steel are also used. Buy heavy-gauge metal and make sure it is prepainted, since unfinished gutters are difficult to paint and to keep from peeling.

Vinyl and wood gutters are also available, but are considerably more expensive. Vinyl is most often found with vinyl siding, and wood with wood roofs or shake siding. If buying wood gutters, get insect- and weather-resistant wood such as heartwood redwood, red cedar or cypress. Otherwise, treat the wood with preservative.

Guttering comes in a variety of patterns. Half-round, rectangular and the popular "OG" patterns are available with matching downspouts. Downspouts are usually corrugated to better resist ice dams.

Installation. Metal gutters are hung either on hangers or on 7-inch spikes inside ferrules. Wood gutters are attached directly to the fascia boards with galvanized screws. All gutters should be located so that the drip edge of the bottom row of shingles flows into the center of the gutter. Use furring strips or wood blocks, with flashing, if necessary. Fasteners should be on 30-inch centers.

The general rule for gutter pitch is ⅛ inch per foot toward the downspout, but you may have to adjust for less in long systems. When gutters are over 35 feet long, it is best to have a downspout at each end, pitching the gutter from the center to each leader.

Fasten the downspout to the wall with straps at the top and bottom, and at six-foot intervals. Shim out with wood blocks or spacers if necessary, as recommended by the manufacturer. Use an elbow to connect the downspout to the gutter at the top. Place another elbow at the bottom to direct the water onto a splash block; this will not be necessary if the downspout connects to a drainage system. In that case, the leader empties directly into the system through a vitrified tile line. Elbow connections are made with slip joints, rivets or solder as recommended by the manufacturer.

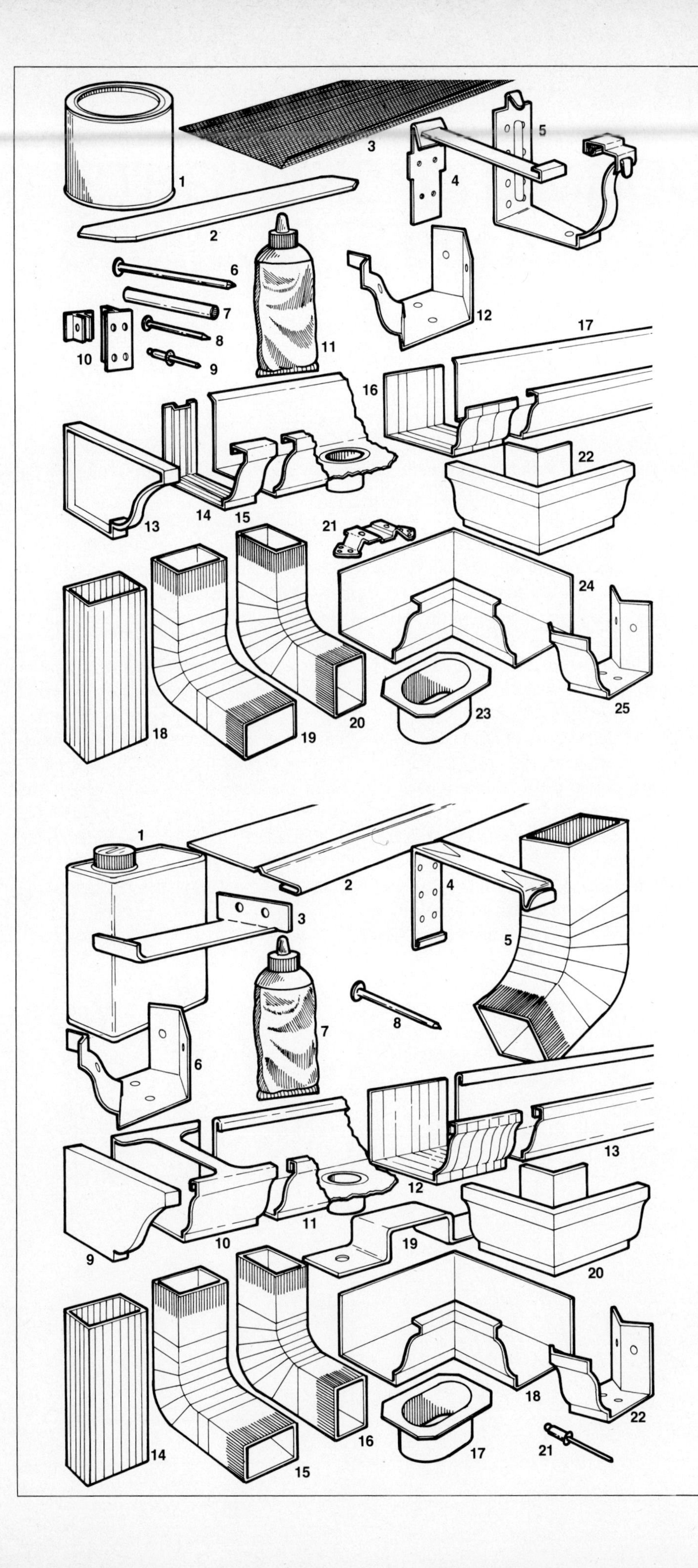

Aluminum gutter system

1 Touch up paint
2 Downspout pipe band
3 Gutter cover
4 Fascia bracket
5 Fascia bracket with clip
6 Spike
7 Ferrule
8 Aluminum nail
9 Pop rivet
10 Downspout pipe bracket
11 Joint sealer
12 Inside miter strip
13 End caps
14 Slip joint connector
15 Outlet tub section
16 Expansion joint
17 Gutter
18 Downspout
19 Square elbow
20 75° Elbow
21 Downspout pipe cleat
22 Outside miter section
23 Outlet tube
24 Inside miter section
25 Outside miter strip

Solid vinyl gutter system

1 PVC surface primer
2 Fascia apron hanger
3 Gutter spacer
4 Gutter hanger
5 Side elbow
6 Aluminum nail
7 PVC cement
8 Inside molded corner
9 End cap
10 Gutter connector
11 Collector outlet
12 Expansion joint
13 Gutter
14 Downspout
15 Square elbow
16 90° elbow
17 Outlet tube
18 Outside miter section
19 Downspout strap
20 Outside molded corner
21 Pop rivet
22 Outside miter strip

11

DORMERS, SKYLIGHTS AND OTHER PROJECTS

DEFINITIONS AND RAFTER SQUARES

Ridge board. The ridge board is the horizontal member used for connecting the upper ends of the rafters on one side to the rafters on the opposite side. In cheap construction the ridge board is usually omitted and the upper ends of the rafters are spiked together.

Common rafters. A common rafter is a roof member extending diagonally from the plate to the ridge.

Hip rafters. A hip rafter is a member extending diagonally from the corner of the plate to the ridge.

Valley rafters. A valley rafter is one extending diagonally from plate to ridge at the line of intersection of two roof surfaces.

Jack rafters. Any rafter that does not extend from plate to ridge is called a jack rafter.

There are different kinds of jacks and according to the position they occupy they may be classified as: hip jacks, valley jacks, and cripple jacks.

Hip jack. A jack rafter with the upper end resting against a hip and lower end against the plate is called a hip jack.

Valley jack. A jack with the upper end resting against the ridge board and lower end against the valley is called a valley jack.

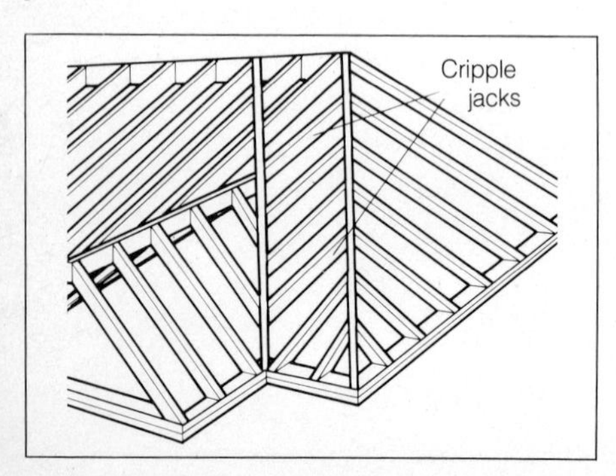

Cripple jack. A jack that is cut in between a hip and valley rafter is called a cripple jack. A cripple jack touches neither the plate nor the ridge.

Rafter cut. All rafters must be cut to proper angles so that they will fit at the points where they are framed.

Top or plumb cut. The cut of the rafter end which rests against the ridge board or against the opposite rafter is called the top or plumb cut.

Bottom or heel cut. The cut of the rafter end that rests against the plate is called the bottom or heel cut. The bottom cut is also called the foot or seat cut.

Side cuts. Besides having top and bottom cuts, hip and valley rafters as well as all jacks must also have their sides at the end cut to a proper bevel so that they will fit into the other members to which they are to be framed. These are called side cuts or cheek cuts.

How to Use a Rafter Square

Although rafters are often cut by the trial-and-error method, as discussed later, the fastest and easiest cutting technique calls for the use of a rafter square. This tool enables you to cut the angles while on the ground rather than measuring and fitting them exactly while sitting on the roofing.

Plumb cut
Side cut
Ridge board
Heel cut
Plate
Jack rafter
Heel cut
Common rafter

The rafter square looks like a carpenter's framing square, but tables have been etched on it to give lengths and angles for cuts. (The material presented here is adapted from information provided by Stanley Tools.)

Because the length of a common rafter is the shortest distance between the outer edge of the plate and a point on the center line of the ridge, this length is taken along the "measuring line", which runs parallel to the edge of the rafter and is the "hypotenuse" or the longest side of a right triangle, the other two sides being the run and the rise.

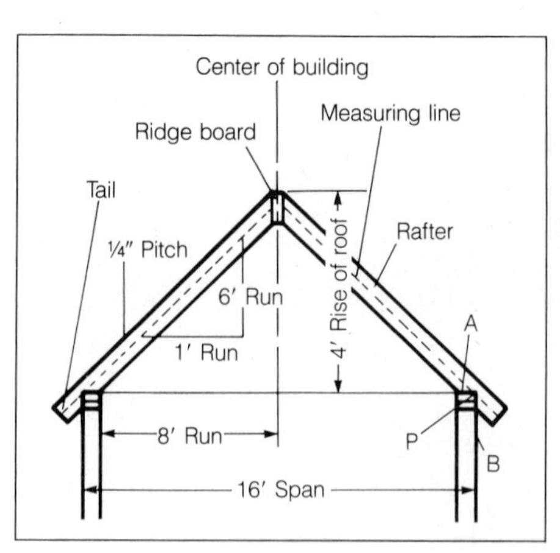

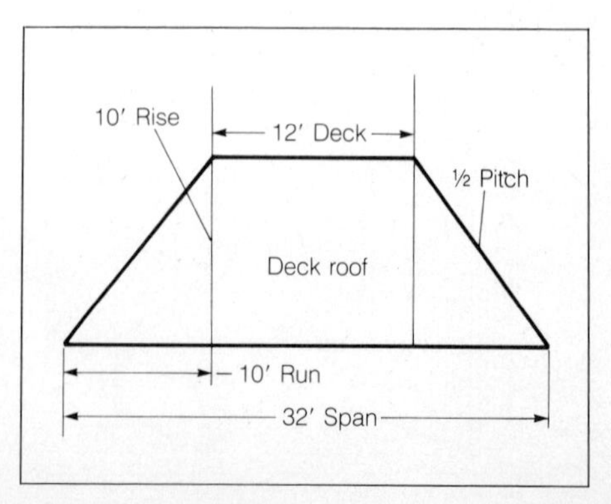

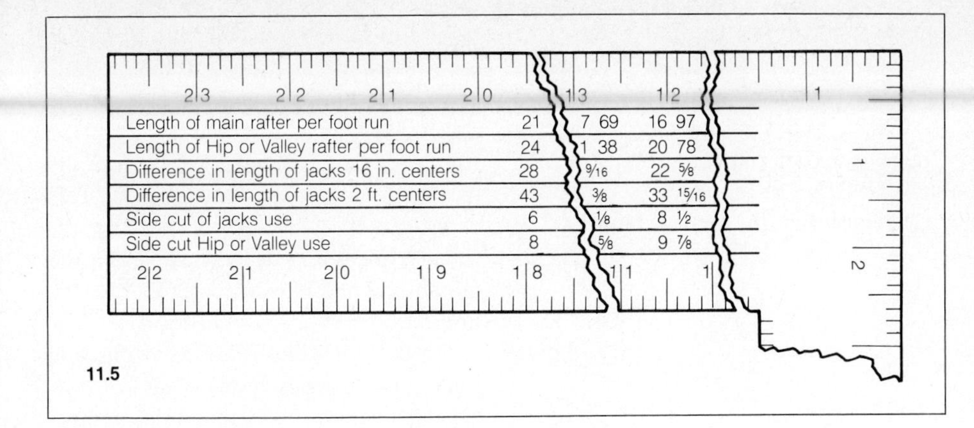

	23	22	21	20	13	12
Length of main rafter per foot run				21	7 69	16 97
Length of Hip or Valley rafter per foot run				24	1 38	20 78
Difference in length of jacks 16 in. centers				28	9/16	22 5/8
Difference in length of jacks 2 ft. centers				43	3/8	33 15/16
Side cut of jacks use				6	1/8	8 1/2
Side cut Hip or Valley use				8	5/8	9 7/8

11.5

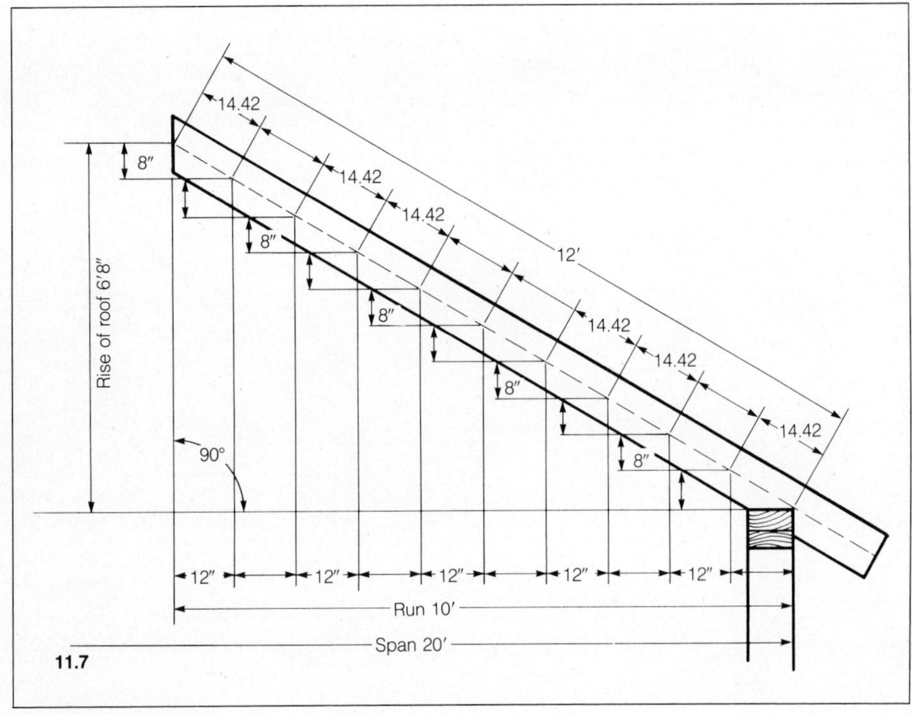

11.7

11.6

The rafter tables on the face of the body include the outside edge graduations on both body and tongue, which are in inches and sixteenths of an inch.

Finding the length of the rafters. The lengths of common rafters are found on the ''first'' line indicated as the length of the main rafters per foot run. There are seventeen of these tables beginning at 2 inches and continuing to 18 inches (see Figure 11.5).

General rule. To find the length of a common rafter—multiply the ''length given in the table'' by the number of feet of the run.

Example. Find the length of a common rafter where the rise of roof is 8 inches per foot run (or one third pitch) and the building is 20 feet wide. First find on the ''inch line'' on the top edge of the body the figure that is equal to the rise of the roof, which in this case will be 8. On the first line under the figure 8 will be found 14.42, which is the length of the rafter in inches ''per foot run'' for this particular pitch (see Figure 11.6).

The building is 20 feet wide. Therefore the run of the rafter will be 20 ÷ 2 (since the run is half the width) equals 10 feet.

Since the length of the rafter per ''one foot run'' equals 14.42 inches, the total length of rafter will be 14.42 multiplied by 10, which equals 144.20 inches (or 144.20 ÷ 12 equals 12.01 feet) for all practical purposes, this means 12 feet (see Figure 11.7).

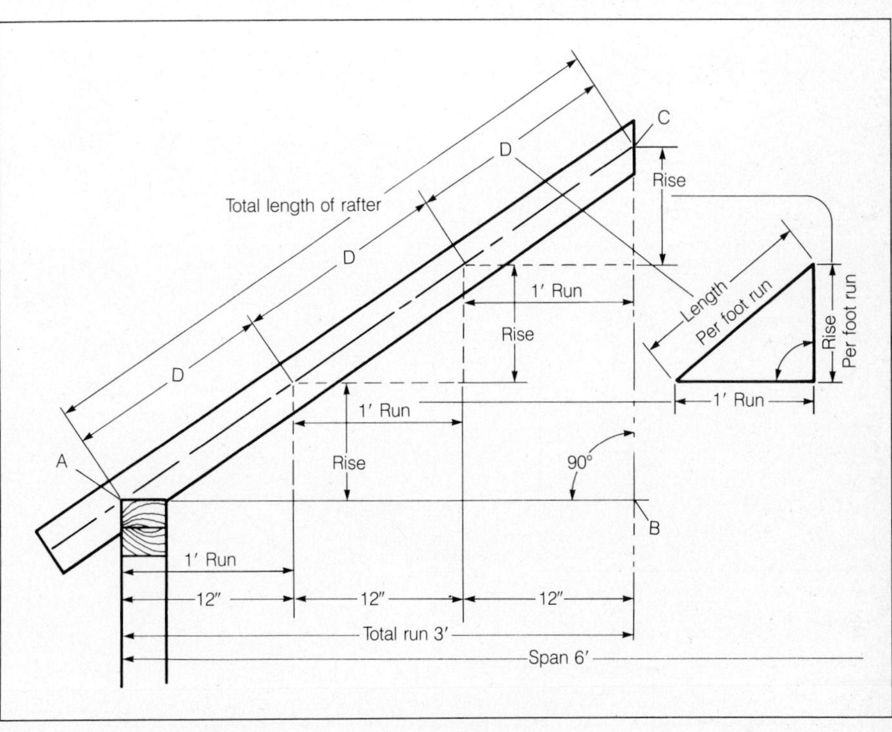

Figuring the top and bottom cuts. The top or plumb cut is the cut at the upper end of the rafter where it rests against the opposite rafter or against the ridge board.

The bottom or heel cut is the cut at the lower end which rests on the plate.

The top cut is parallel to the center line of the roof, the bottom cut is parallel to the horizontal plane of the plates. Therefore, the top and bottom cuts are at right angles to each other.

General rule. To obtain the top and bottom cuts of a common rafter, use 12 inches on the body and the "rise per foot run" on the tongue. 12 inches on the body will give the horizontal cut; the figure on the tongue gives the vertical cut.

Example. To illustrate, we will imagine a large square placed alongside the rafter as shown in Figure 11.8. We may notice that the edge of the tongue coincides with the top cut of the rafter and the edge of the blade coincides with the heel cut. If this square were marked in feet it would show the run of the rafter on the body and the total rise on the tongue. The line "AB" would give us the bottom cut and line "AC" the top cut.

However, the regular square is marked in inches and since the relation of the rise to one foot run is the same as the total rise bears to the total run, we use 12 inches on the blade, and the "rise per foot" on the tongue to obtain the respective cuts. The distance "12" is used as a unit and is the "one foot run" while the figure on the other arm of the square represents the "rise" per foot run (see Figures 11.8A and 11.8B).

Determining the actual length. The lengths of rafters obtained from the tables are "to the center line of the ridge." Therefore, the thickness of half of the ridge board should always be deducted from the obtained total length before the top cut is made (see Figure 11.9). This deduction of half the thickness of the ridge is measured at right angles to the plumb line and is marked parallel to this line.

Figure 11.10 illustrates the wrong and right ways to measure the length of rafters. The diagram "D" shows the measuring line as the edge of the rafter which is the case when there is no tail or eave.

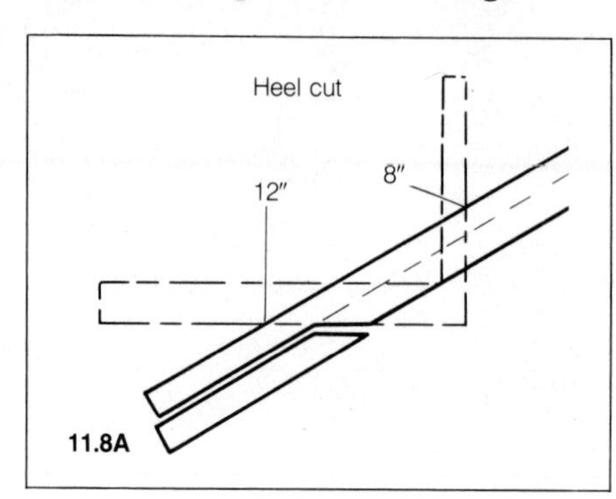

11.8A

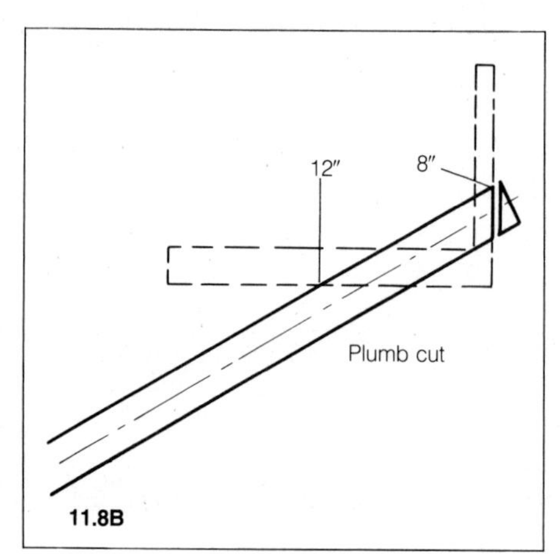

11.8B

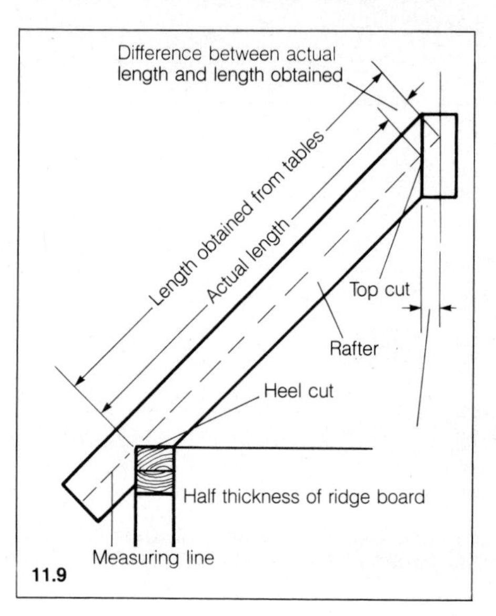

11.9

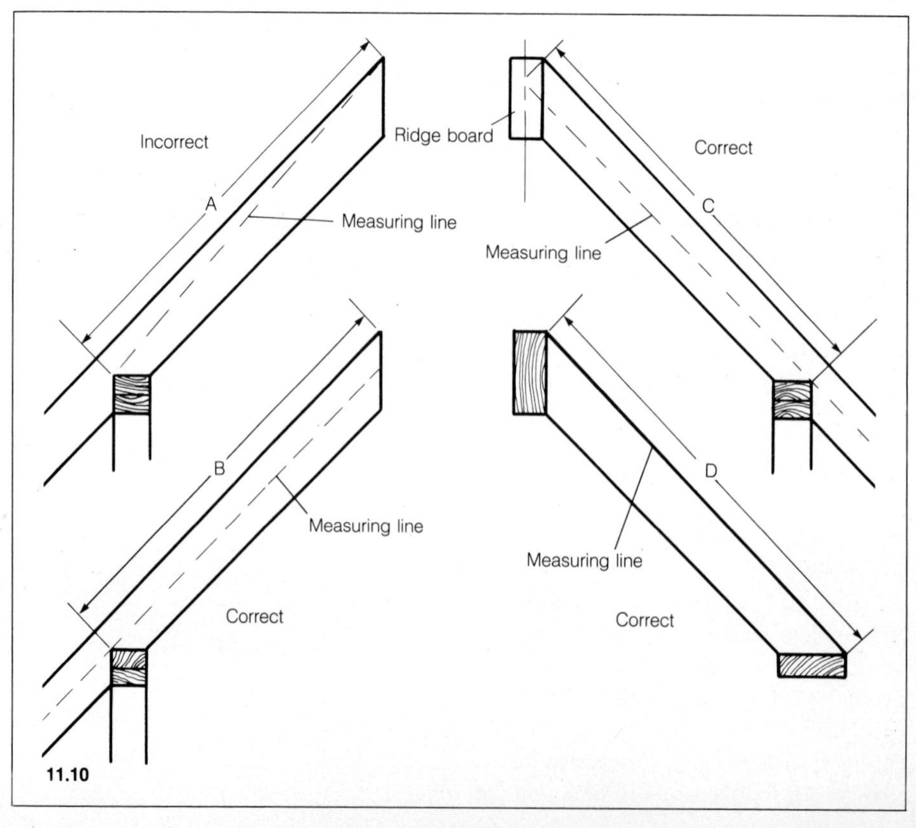

11.10

Applying the square. After the total length of the rafter has been established, both ends should be marked and allowance made for a tail or eave, and for half the thickness of the ridge.

Both cuts are obtained by applying the square so that the 12-inch mark on the body, and the mark on the tongue that represents the rise, are at the edge of the stock.

All cuts for common rafters are made at right angles to the side of the rafter.

Example. A common rafter is 12 ft. 6 inches, the rise per foot run being 9 inches. Obtain the top and bottom cuts (see Figure 11.11).

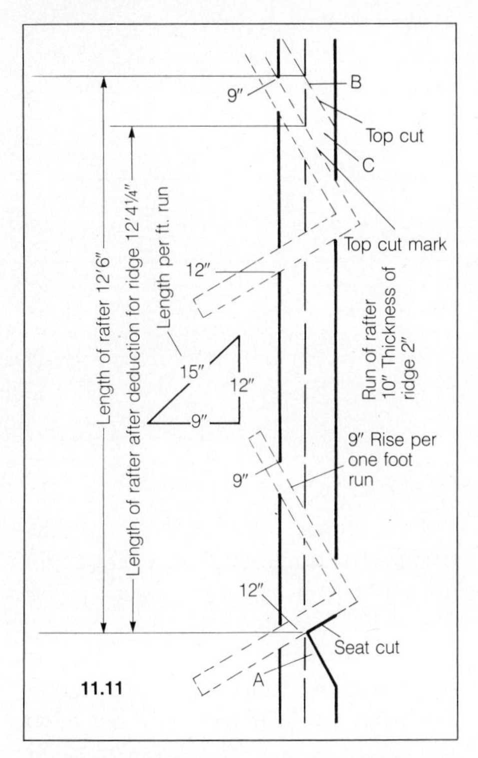

11.11

Points "A" and "B" are the ends of the rafter. To obtain the bottom or seat cut take 12 inches on the body of the square and 9 inches on the tongue. Lay the square on the rafter so that the body will coincide with point "A" or the lower end of the rafter. Mark along the body of the square and cut.

To obtain the top cut move the square so that the tongue coincides with point "B" which is the upper end of the rafter. Mark along the tongue of the square.

Deducting for the ridge. The deduction for half the thickness of the ridge should now be measured. Half the thickness of the ridge is 1 inch. One inch is deducted at right angles to the top cut mark or plumb line, point "C." A line is then drawn parallel to the top cut mark and the cut made. You will notice that the allowance for half the ridge measured along the measuring line is 1¼ inches. This will vary according to the rise per foot run. It is therefore important to measure for this deduction at right angles to the top cut mark or plumb line.

How to measure rafters. The length of rafters having a tail or eave can also be measured along the back or top edge instead of the measuring line, as illustrated in Figure 11.12. To do this, it is necessary to carry a plumb line to the top edge from P and the measurement started from this point.

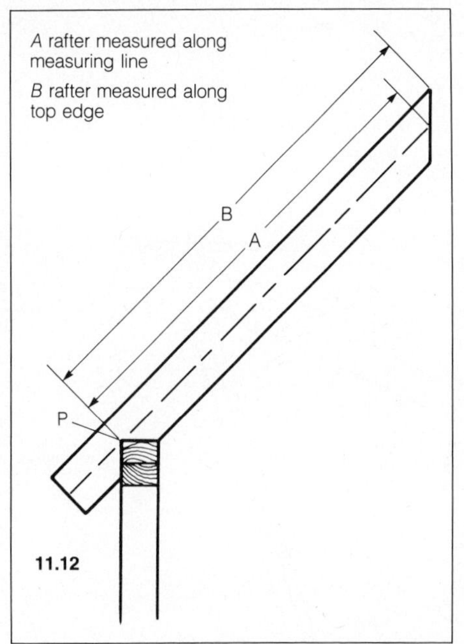

11.12

Adding odd inches of common rafters. Occasionally in framing a roof the run may have an odd number of inches as in the case of a building with a span of 24 feet 10 inches. This would mean a run of 12 feet 5 inches. The additional five inches can be easily added without mathematical division after the figures obtained from the square for 12 feet of run are measured. The additional five inches are measured at right angles to the last plumb line, as illustrated in Figure 11.13.

Figuring Hip and Valley Rafters

The hip rafter is a roof member that forms a "hip" in the roof, usually extending from the corner of the building diagonally to the ridge.

The valley rafter is similar to the hip only that it forms a "valley" or depression in the roof instead of a hip. It also extends diagonally from plate to ridge. Therefore the total rise of hip and valley

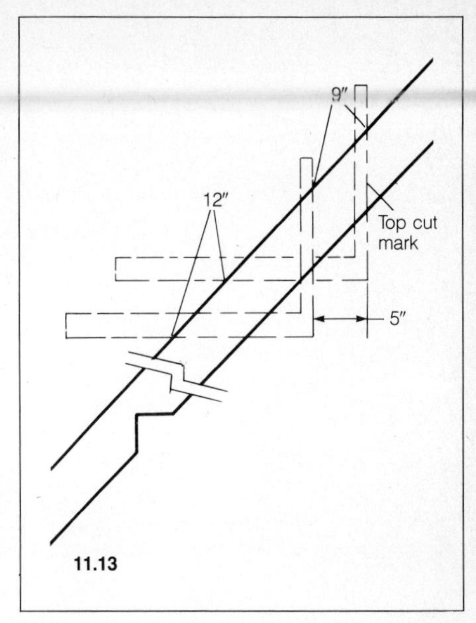

11.13

rafters is the same as that of common rafters.

The relation of hip and valley rafters to common rafters is the same as the relation of the sides of a right triangle; therefore, it will be well to explain here one of the main features of right triangles: in a right triangle if the sides forming the "right angle" are 12 inches each, the hypotenuse or the side opposite the right angle is equal to 16.97 inches which is usually taken as "17" inches (see Figure 11.15).

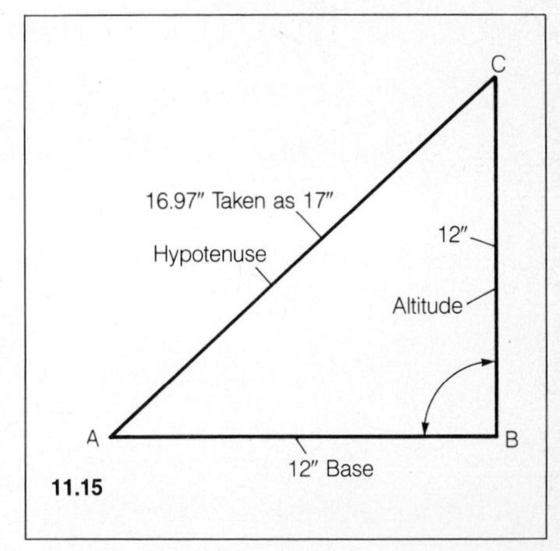

11.15

The position of the hip rafter and its relation to the common rafter is plainly illustrated in Figures 11.16 and 11.18 and where the hip rafter is compared to the "diagonal" of a square prism.

The prism has a base of 5 feet square and its height is 3 feet 4 inches.

"D" is the corner of the building,
"BC" is the total rise of the roof,
"AB" is the run of the common rafter,
"AC" is the common rafter,
"DB" is the run of the hip rafter,
"DC" is the hip rafter.

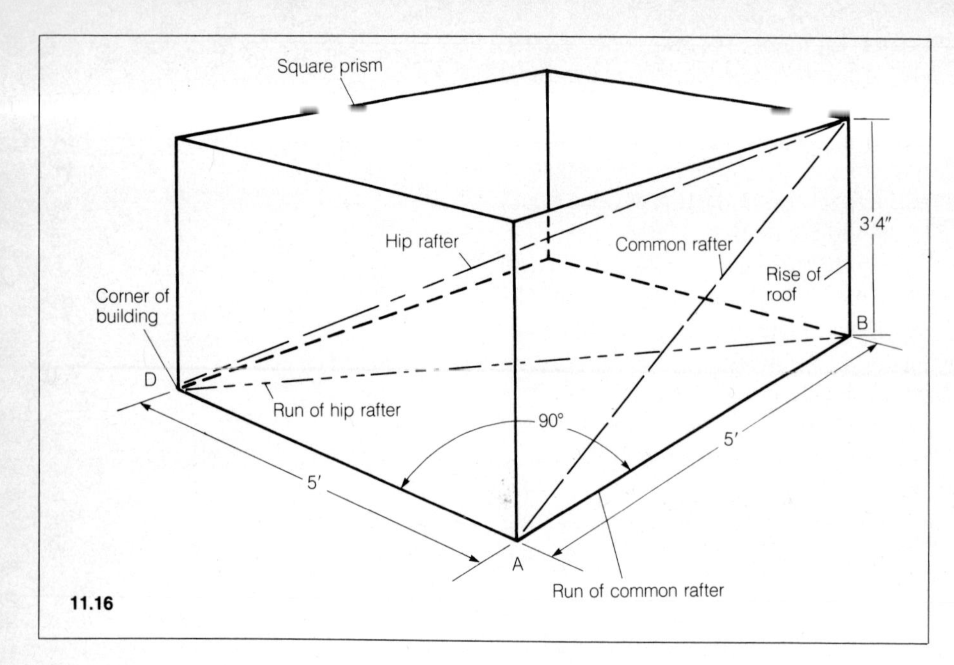

11.16

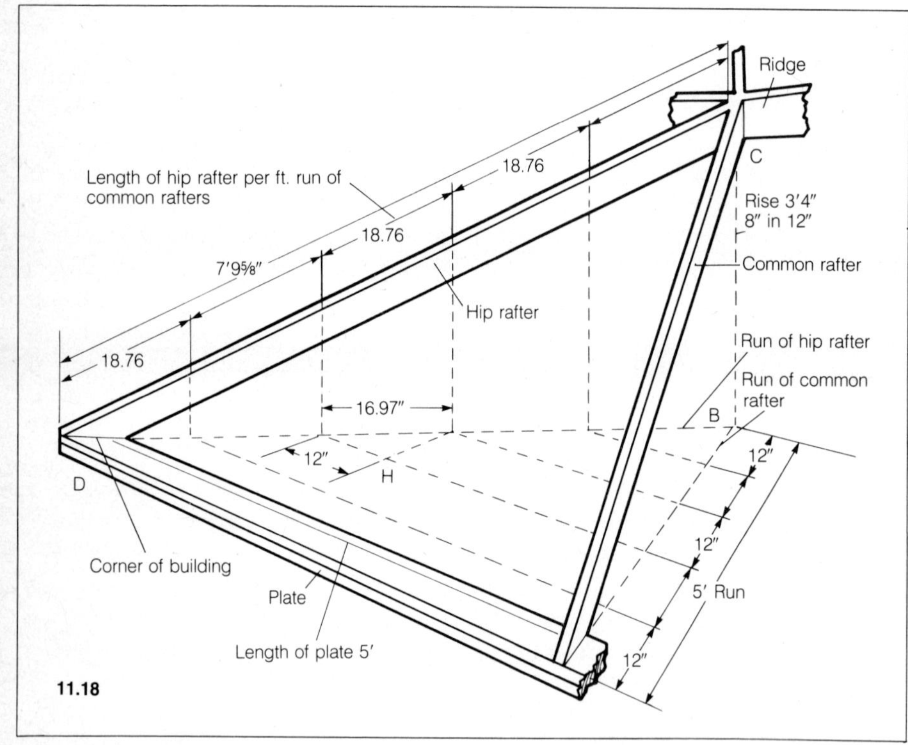

11.18

It will be noted that the figure "DAB" is a right triangle whose sides are: the portion of the plate—"DA," the run of common rafter—"AB" and the run of hip rafter—"DB." The run of the hip rafter being opposite the right angle "A" is the "hypotenuse" or the longest side of the right triangle.

If we should take only one foot of run of common rafter and one foot length of plate we will have a right triangle "H" whose sides are each 12 inches long and whose hypotenuse is 17 inches or more accurately 16.97 inches (see Figure 11.18).

The hypotenuse in this small triangle "H" is a portion of the run of the hip rafter "DB," which corresponds to one foot run of common rafter.

Therefore, the "run of hip rafter" is always 16.97 inches for every 12 inches of foot run of common rafter, and the "total run" of hip rafter will be 16.97 inches multiplied by the number of feet run of common rafter.

Lengths of hip and valley rafters. The lengths of hip and valley rafters are found on the "second" line of the rafter table entitled: "Length of hip or valley rafters per foot run," which means that the figures in the table indicate the length of hip and valley rafters "per foot run of common rafters" (see Figure 11.5).

General rule. To find the length of a hip or valley rafter—multiply the length given in the table by the number of feet of the run of common rafter.

Example. Find the length of a hip rafter where the rise of roof is 8 inches per foot run or one third pitch and building is 10 feet wide (see Figure 11).

Proceed the same as in the case of common rafters, i.e., find on the "inch line" of the body of the square the figure corresponding to the rise of roof—which is 8. On the "second" line under this figure is found "18.76" which is the length of hip rafter in inches for each foot of run of common rafter for one third pitch (see Figure 11.17).

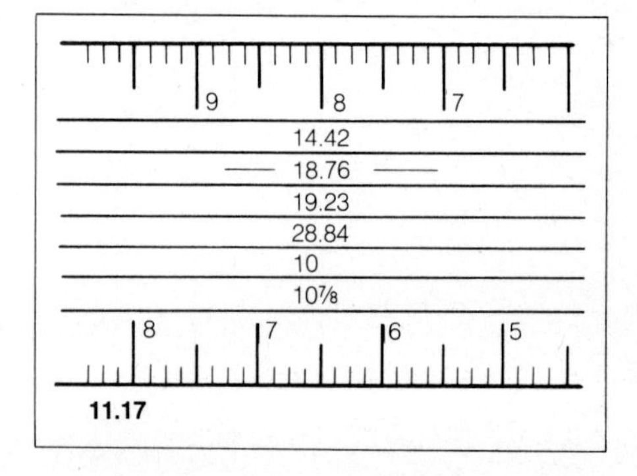

11.17

The common rafter has a 5 foot run and therefore there are also 5 equal lengths for the hip rafter, as may be seen in the illustration (see Figure 11.18).

We have found the length of the hip rafter to be 18.76 inches per one foot run. Therefore the total length of hip rafter will be 18.76 x 5=93.80 inches=7.81 feet or for practical purposes 7 feet 9¹³⁄₁₆ inches.

Top and bottom cuts. The following rule should be followed for top and bottom cuts.

General rule. To obtain the top and bottom cuts of hip or valley rafters use 17 inches on the body and the "rise per foot run" on the tongue. 17 on the body will give the seat cut and the figure on the tongue the vertical or top cut (see Figure 11.19).

Measuring hip and valley rafters. The length of all hip and valley rafters must always be measured along the center of the top edge or back. The rafters with a tail or eave are treated similar to common rafters as mentioned earlier and as illustrated in Figure 11.12, except the mea-

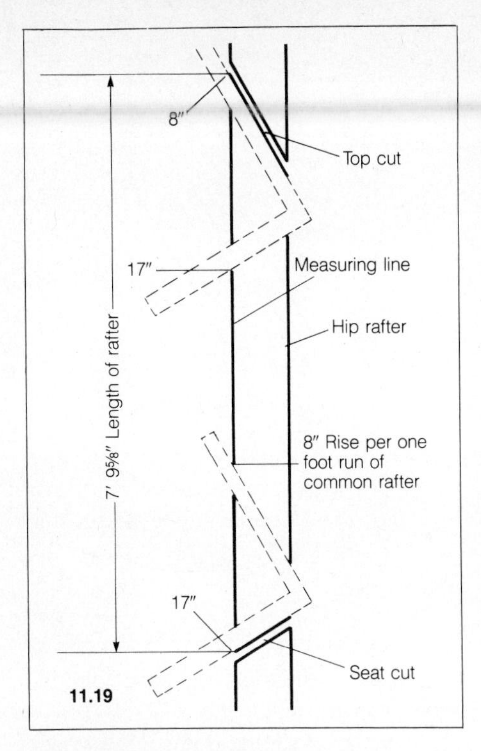

11.19

surement or measuring line is the center of the top edge.

Deducting from hip or valley rafter for ridge. The deduction for the ridge is measured the same as for the common rafter, Figure 11.11, except that half the diagonal (45°) thickness of the ridge must be used.

Side cuts. Hip and valley rafters in addition to the top and bottom cuts must also have side or cheek cuts at the point where they meet the ridge.

These side cuts are found on the "sixth" or bottom line of the rafter tables which is marked: "Side cut hip or valley—use."

The figures given in this line refer to the graduation marks on the "outside edge of the body" (see Figure 11.5).

The figures on the square have been derived by determining the figure to be used with 12 on the tongue for the side cuts of the various pitches by the following method. From a plumb line the thickness of the rafter is measured and marked at right angles as at A, Figure 11.20. A line is then squared across the top of the rafter and the diagonal points connected as at B. The line B or side cut is obtained by marking along the tongue of the square.

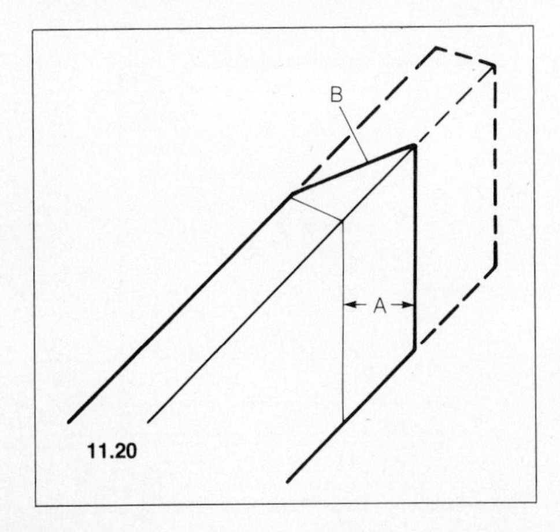

11.20

General rule. To obtain the side cut for hip or valley rafters—take the figure given in the table—on the body of the square and 12 inches on the tongue. Mark side cut along the tongue where the tongue coincides with the point on the measuring line.

Example. Find side cut for hip rafter—the roof having 8 inches rise per foot run or one third pitch (see Figures 11.21 and 11.22).

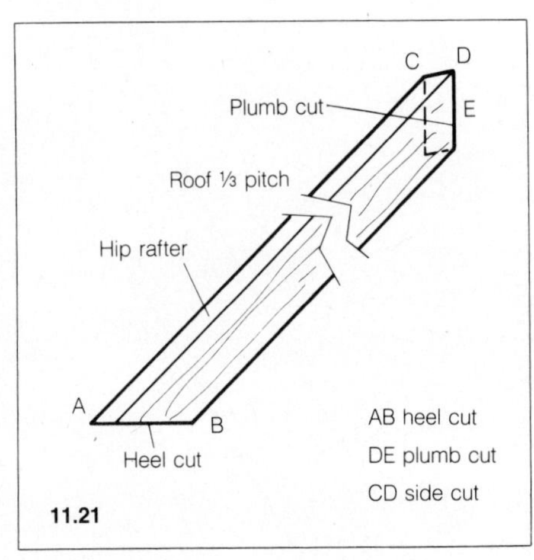

11.21

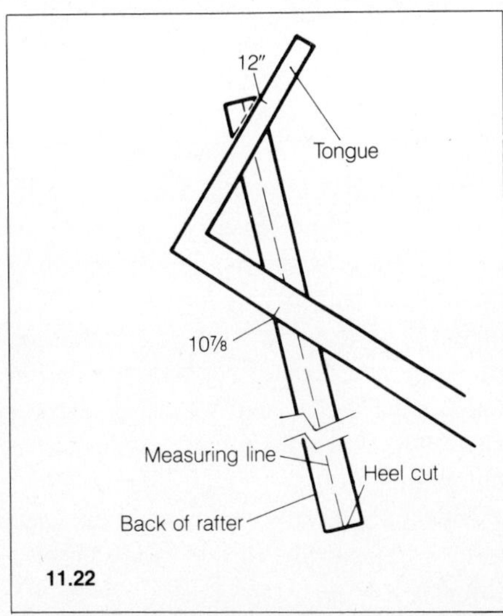

11.22

Figure 11.21 represents the position of the hip rafter on the roof. The rise of roof being 8 inches to the foot, first locate the figure 8 on the outside edge of the body. Under this figure in the bottom line you find "10⅞." This figure is taken on the body and 12 inches on the tongue. The square is applied to the edge of the back of the hip rafter. The side cut "CD" comes along the tongue.

The deduction for half the thickness of the ridge must be determined and measured the same as for the common rafters, Figure 11.11, except that half the diagonal (45°) thickness of the ridge must be used.

In making the seat cut for the hip rafter an allowance must be made for the top edges of the rafter which would project above the line of the common and jack rafters if the corners of the hip were not removed or "backed." The hip rafter must be slightly lowered by cutting parallel to the seat cut a distance which varies with the thickness and pitch of the roof.

It should be noted that on the square shown the 12 inch mark on the tongue is always used in all angle cuts, both top, bottom and side, thus leaving the workman but one number to remember when laying out side or angle cuts, namely the figure taken from the fifth or sixth line in the table.

The side cuts come always on the "right hand" or "tongue" side on rafters. When marking boards these can be reserved for convenience at any time by taking the 12 inch mark on the body and using the body references on the tongue.

Odd inches of hip and valley rafters. To obtain additional inches in run of hip or valley rafters similar to the explanation for common rafters (Figure 11.13), the diagonal (45°) of the additional inches or approximtely 7¹⁄₁₆ inches for five inches of run should be used in a similar manner.

Jack Rafters

Jack rafters are "discontinued" common rafters—or common rafters "cut off" by the intersection of a hip or valley before reaching the full length from plate to ridge.

Jack rafters lie in the same plane with common rafters. They usually are spaced the same and have the same pitch and therefore they also have the same length per foot run as common rafters have.

Jack rafters are usually spaced 16 inches or 24 inches apart and as they rest against the hip or valley equally spaced the second jack must be twice as long as the first one, the third three times as long as the first and so on (see Figure 11.23).

Length of jack rafters. The lengths of

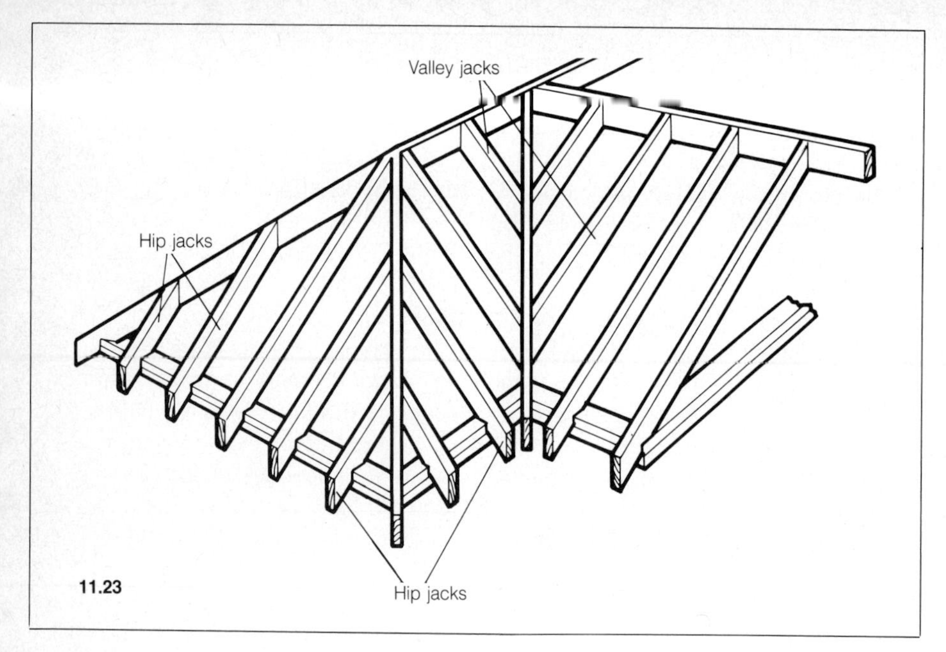

11.23

11.25

jacks are given in the third and fourth line of the rafter tables and are indicated by:

(3rd line) "Difference in length of jacks—16 inches centers";

(4th line) "Difference in length of jacks—2 feet centers."

The figures in the table indicate the "length of the first or shortest jack" which is also the difference in length between the first and second, between the second and third jack and so on.

General rule. To find the length of a jack rafter, multiply the value given in the tables by the number indicating the position of the jack. From the obtained length subtract half the diagonal (45°) thickness of the hip or valley rafter.

Example. Find the length of the second jack rafter, the roof having a rise of 8 inches to 1 foot of run of common rafter, the spacing of jacks being 16 inches.

On the outer edge of the body find figure 8 which corresponds to the rise of roof. On the third line under this figure find "19¼." This means that the first jack rafter will be 19¼ inches long. Since the length of the second jack is required, multiply 19¼ by 2 which equals 38½ inches. From this length half the diagonal (45°) thickness of the hip or valley rafter should be deducted in the same manner as the deduction was made on the hip rafter for the ridge.

Proceed in the same manner when you need to know lengths of jacks spaced 24 inch on center. It should be borne in mind that the second jack is twice as long as the first one, the third jack is three times as long as the first one, and so on.

11.24

Top and bottom cuts. Since jack rafters have the same "rise per foot run" as common rafters, the method of obtaining the top and bottom cuts is the same as for common rafters; i.e., take 12 inches on the body and the rise per foot run on the tongue; 12 inches will give the seat cut and the figure on the tongue, which is the plumb cut.

Side cut. At the end where the jack rafter frames to the hip or valley rafter a side cut is required.

The side cuts for jacks are found on the "fifth line" of the rafter tables and which is marked: "Side cut of jacks—use" (see Figure 11.5).

General rule. To obtain the side cut for a jack rafter, take the figure shown in the table—on the body of the square and 12 inches on the tongue. Mark along the tongue for the side cut.

Example. Find the side cut for jack rafters of a roof with 8 inch rise per foot run or ⅓ pitch (see Figures 11.24 and 11.25). Under the figure 8 in the fifth line of the table find "10." This figure taken on the outside edge of the body—and 12 inches on the tongue will give the required side cut.

Finding Brace Measurements

This table will be found along the center of the back of the tongue and gives the lengths of common braces.

Example. Find the length of a brace whose run on post and beam equals 39 inches (see Figure 11.27). In the brace table find the following expression:

39
39 55.15.

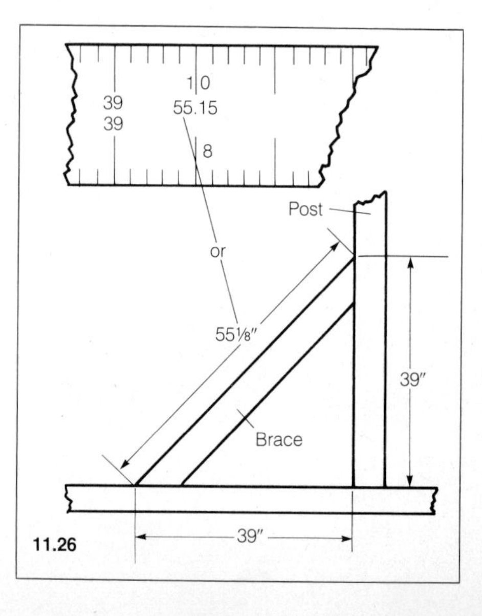

11.26

This means that with a 39 inch run on the beam and a 39 inch run on the post the length of the brace will be 55.15 inches, or for practical purposes 55⅛ inches.

Braces may be regarded as common rafters. Therefore, when the brace run on the post differs from the run on the beam, their lengths as well as top and bottom cuts may be determined from the figures given in the tables of common rafters.

BUILDING DORMERS

One of the best ways to gain additional space is to install a dormer (or, in the case of gable dormers, two or three). This is especially true of one-and-a-half or Cape Cod style homes, where the steeply pitched roof limits ceiling height.

Whereas many additions must be carefully planned to blend in with the architecture of the home, most dormers can be designed and added without greatly affecting the overall look of the home.

Gable vs. Shed Dormers

Gable dormers normally have a single window and a peaked roof, while shed dormers have a flatter, sloping roof and extend the width of a room or of the entire second floor. Shed dormers are not as attractive as gable dormers and should be placed at the back of the home, if possible. Gable dormers, on the other hand, can add to the appearance of a house by breaking up what may be an otherwise uninteresting roofline. The type you choose will depend on the space needed, the planned interior layout, and the architectural style of the house.

Gable dormers. Usually two or more gable dormers are added to a roof plan. Space gable dormers so that they give a pleasing proportion to the home's elevation. When there are two dormers, they usually are set equidistant from the ends of the roof. When there are more than two, intermediate dormers should be equally spaced between the outer ones for a balanced look. The dormers can be built out to the roof edge or set back into the roof, but they all should be the same depth.

Shed dormers. Shed dormers can be any size you wish, extending the full width of the house and up to the eaves if desired. They usually look better set at least slightly back from the roof edge, especially if the corner is narrow. Many times, shed dormers reach all the way to

A small gable dormer can be added to a roof without significantly changing the appearance of the home. The window will provide a source of light and ventilation for an attic room.

A shed dormer will significantly increase the usable living area on a second floor. This small home was converted to a two-family with the addition of the shed dormer.

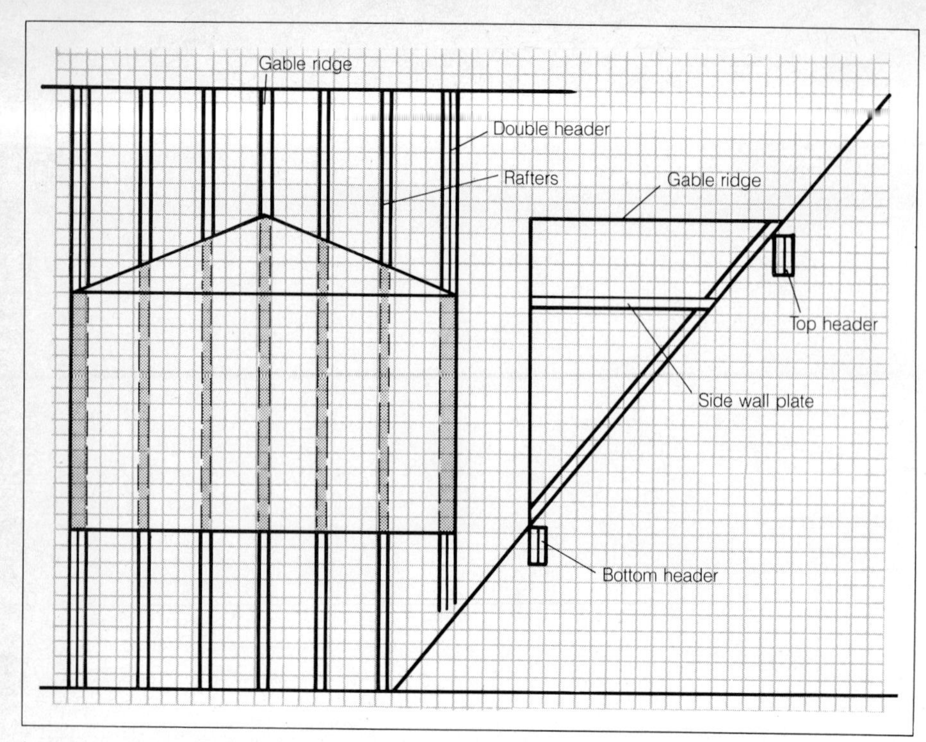

In order to plan properly, you must prepare a scale drawing of the proposed dormer, showing all rafters to be cut and the position of new framing.

the roof peak. This adds headroom. However, the closer you come to the peak, the narrower a shed dormer becomes. You can't build it wider than the ridge. For a hip roof, gable dormers usually are preferred to shed dormers, unless you have a wide ridge board.

Drawings. Once you have chosen the type and location of your dormer(s), work up a scaled drawing or set of drawings to show all the framing. Determine the location of all the existing framing members in the area where the dormer will be located, and where you want to attach the new framing. Construction details, lumber specifications and spacing may be designated in your local code, so look up code requirements before you draw your plans.

Using graph paper and a scale of ¼ inch equals one foot, draw a representation of the front of the dormer; also draw the profile of the dormer with the slope of the roof. On the side view, mark the lower edge of the top header and then add one sidewall plate and the bottom header. Mark the spot where the gable ridge intersects the roof. On the front view, draw in the doubled rafters. Then add horizontal lines to indicate the headers. To represent the gable ridge of the dormer, add a dotted line.

Materials. For the walls of the dormer, use 2x4s set on 16-inch centers. Usually 2x4s are acceptable for gable roofs, but check local building codes. Most shed dormers use 2x6s for rafters, but larger spans may require 2x8s. Consult local building codes for this specification; a good rule of thumb is to use the same size rafters as the existing ones. The new roof should have a pitch of at least two inches per foot. This enables you to use the same materials on the roof that you now have. A flatter pitch will require roll roofing.

The dormer walls should be faced in the same materials as the rest of the house. Use the drawing to estimate the amount of sheathing, siding, decking, and other materials you require, and have them on hand before you begin.

Codes. As mentioned above, the building code may specify the size of framing you must use, among other things. It is a good idea to check the local building code even before you begin to plan. The code may contain some unpleasant surprises, such as forbidding dormer construction (this is not likely). You may also have to obtain a building permit and be subject to inspections before you are finished.

How to Build a Gable Dormer

Whenever possible, plan the gable dormer so that its sides are located at existing rafters. If this cannot be arranged, nail additional rafters at the edges of the cut opening. This should be done before cutting any of the roof openings. The new rafters toenail to the wall plate and to the ridge beam.

Step one: marking the opening. Mark the outline of the opening on the underside of the deck, inside the attic. Drive nails through all four corners and use these as guidelines to also mark the exterior roof shingles for the opening. Snap chalklines between the corners indicated by the nails.

Step two: cutting the opening. From the inside, use a keyhole saw to cut the first 6 inches. Finish cutting the decking

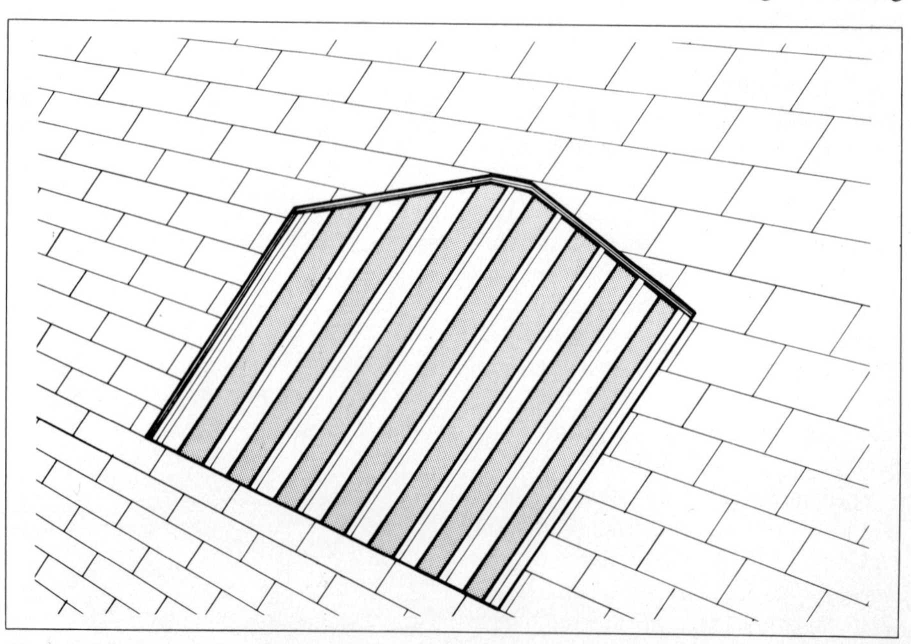

After carefully locating the proper position of the new dormer, you must remove the shingles and roof sheathing to expose the rafters that are to be removed.

from the outside with a reciprocating saw, saber saw, crosscut saw, or circular saw. You must be careful to cut to the depth of the shingles and deck only; do not cut through the rafters. Whatever saw you use, be very careful. Wear goggles and watch out for nails in the deck.

Remove the cut-out section of the deck and roofing by tapping the area, from the inside, with a hammer or by prying it out with a crowbar. Save the shingles from this area for possible later use. Wear a hardhat while working in the attic and watch out for nails protruding down through the deck.

Step three: cutting the rafters. Before you can cut any rafters you must make arrangements to preserve the structural strength of the roof framing. If cutting any more than one rafter, use temporary braces up near the ridge or use a post beneath the ridge beam, centered on the dormer cutout. If possible, position the post over a load-bearing wall. The main bearing wall in a house usually can be found below, in line with the ridge beam. To create the post, nail two 2x4s together. Wedge them between the ridge above and the joist below.

Start the rafter cut at the top of the opening. Cut at right angles to the long-side of the dormer. Now cut the rafters at the bottom of the opening, sawing the rafters plumb (on a true vertical). Keep the faces of all the rafters in the same plane. A string guide is a good idea.

Step four: framing the opening. For the header, use lumber that is the same size as the rafters. Nail it between the flanking rafters at the opening's upper edge. Then nail through the header into the sawed-off ends of the intermediate rafters. Fasten a second header to the first, nailing it also to the flanking rafters. This creates a double header.

Between the flanking rafters at the bottom of the opening, nail another header. Nail into the ends of the intermediate rafters. Use 16d common nails throughout. Since the bottom header serves as a sill for the window framing, it must also be doubled. Nail a second length of lumber to the first bottom header, so that you now have double headers at the top and bottom of the opening.

Using 12d common nails, fasten additional rafters to the flanking rafters at each side of the opening. Toenail the rafters to the ridge board and to the top plate of the wall, cutting the lumber to fit the angles. These need not extend further than the top plate. Then remove about 2 inches of the shingle material all around the opening.

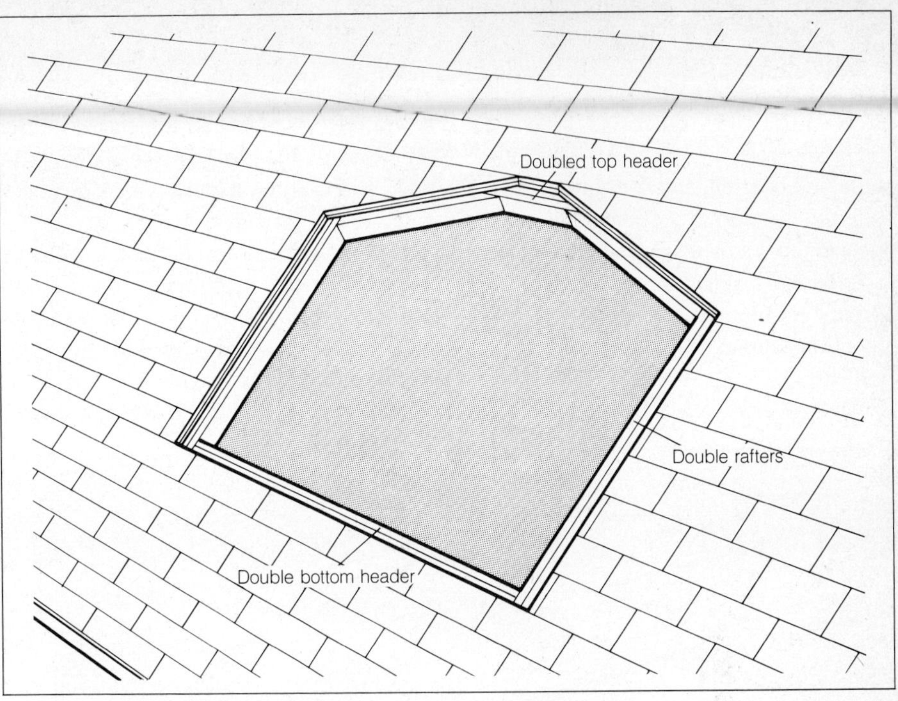

The entire opening must be double-framed to provide a strong perimeter for the additional framing to be attached to the roof. Use doubled side rafters and headers.

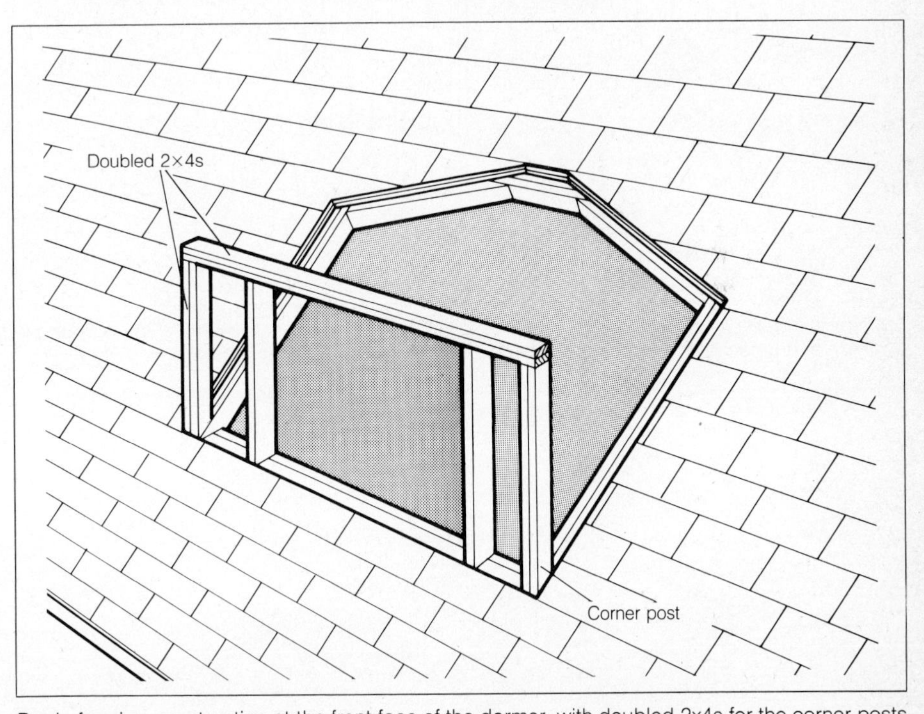

Begin framing construction at the front face of the dormer, with doubled 2x4s for the corner posts and header. Interior studs are positioned to frame the window installation.

Step five: creating an upper sill. Frame in the triangle created by the point where the ridge beam meets the roof slope and the two outer corners of the top double header of the opening.

To do this, drive a nail at the point where the ridge meets the roof. The nail will be centered over the double header. Cut the side of the triangle by resting a 2x4 against the nail and the outer corner of the doubled top header. Mark the 2x4 and cut to fit. Repeat for the other side of the triangle. Remove the nail you used as a reference point. To create the upper sill, nail the two sides of the triangle into the rafters in the existing roof below.

Step six: framing the dormer. Make

corner posts of doubled 2x4s, cutting to fit the bottom. Find the correct angle at the bottom with a protractor, or use a rafter or framing square, or follow the trial-and-error fit method. This same angle will be used later for bottoms of the studs of the dormer. Toenail the corner posts to the doubled side rafters and nail a doubled 2x4 header on top. Check that all corners are plumb. The top header board should be mitered 45 degrees on its narrow side.

Cut two 2x4s to run horizontally from the front of the dormer back to the roof and to serve as top plates for the sides of the gable. The fronts of these 2x4s should be mitered to meet the top board of the top plate of the front of the gable. The other end is mitered on its wide side to the angle of the rafters.

Cut and nail the studs for the sides of the dormer front, allowing for window framing in front. Use 16d nails for nailing into end grains and 12d nails for toenailing throughout.

Step seven: placing the ridge beam. Cut the dormer ridge beam from 2x4 or 2x6 lumber. Angle the back of the ridge beam to fit the top of the upper sill. Cut a center stud for the front of the dormer and let the ridge beam rest on this stud. The beam should be perfectly level. Nail the stud perpendicular to the front wall top plate. Nail the front end of the ridge beam to this stud. Toenail them with 10d nails. Check for a true horizontal with a spirit level. Adjust with shims or by moving the ridge. Measure, cut and fasten the two front rafters of the dormer to help lock the ridge in place.

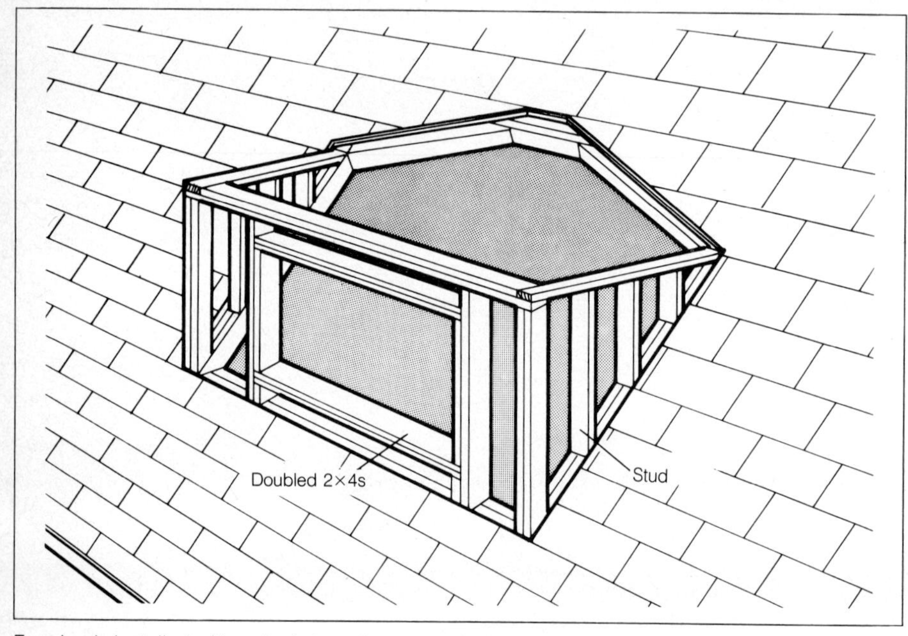

Framing is installed with a single top plate on each side. Install studs with the wide face toward the front of the dormer. Notch top board of double header to fit the side plates.

Step eight: installing the rafters. Nail studs between the front dormer rafters and the top dormer plate. Measure and cut the valley rafters, either using trial and error for the compound angles, or by using a framing square. (A compound angle of a rafter refers to a case in which a piece of lumber has more than one face that must be cut to fit in order to meet other, adjoining angles.) Nail the valley rafters to the doubled flanking rafters, side dormer plates, ridge beam and top header. Use at least 12d nails. Now install the rest of the dormer rafters, spacing them every 16 inches. Cut each jack rafter individually. They will run between the valley rafters and the roof's ridge beam. The dormer is now structurally complete, and the temporary support posts may be removed.

Step nine: adding ceiling joists. If your gable is a small one, you probably will not want ceiling joists inside, since they would cut down on the apparent size of the interior. Larger dormers may have enough headroom for them, in which case you might choose to run 2x4 joists between the sides of the dormer to form a ceiling. Toenail them to the side studs and the top dormer plates.

Step ten: replacing the sheathing and shingles. Once the structural work has been completed, nail plywood or 1x6 sheathing to the roof and walls of the dormer. Fill in the scarred portion of the roof, where you cut through the original sheathing and shingles. Patch the deck and the shingles using pieces of the original sheathing and shingles. You may have to provide "nailers," or scrap pieces of wood, to the rafters or the corners of the dormer to provide nailing surfaces for the finishing materials.

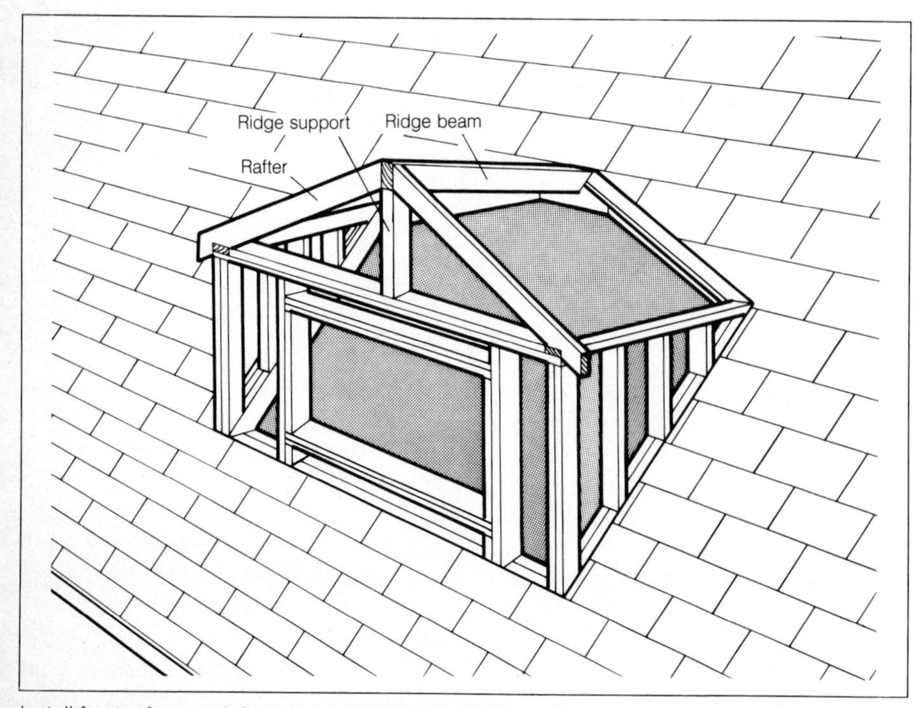

Install front rafters and ridge beam for roof framing. The rafters must be cut to fit the angle of the ridge and notched to meet and fit over the side top plates.

Step eleven: flashing and finishing. Apply apron flashing, starting at the base

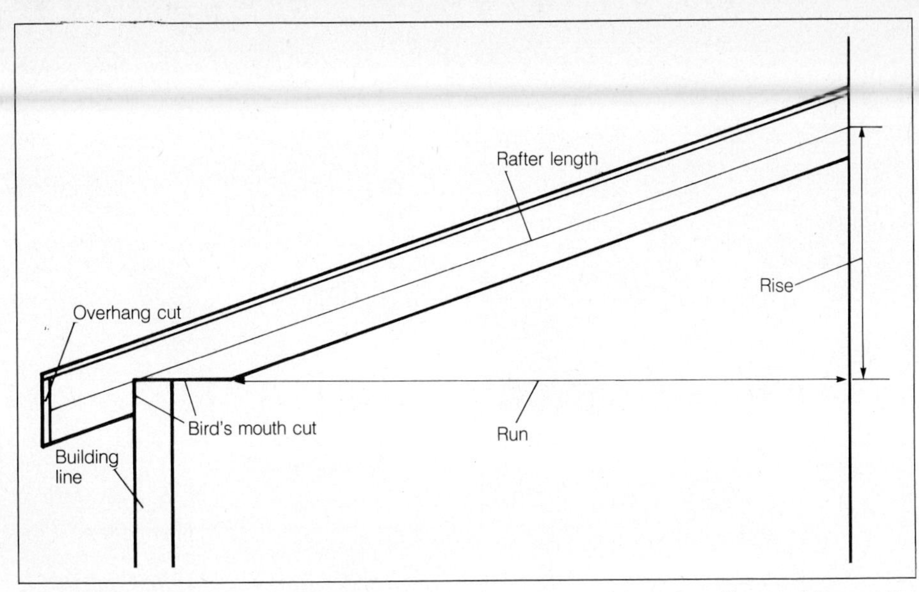

The notch on the rafter is called a "bird's mouth". It is cut so that the rafter will extend beyond the front top plate and attach securely to the front framing.

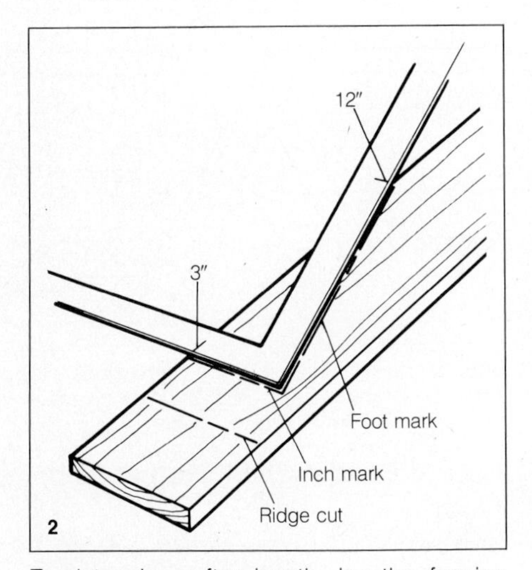

To determine rafter length, lay the framing square along the board with the rise per foot on the tongue and the foot mark on the long leg, aligned with the edge.

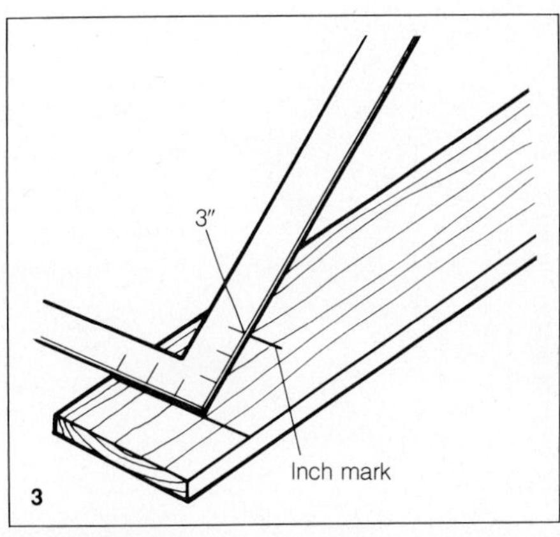

Then align the tongue of the square with the mark for the ridge. Mark off the number of inches beyond full number of feet. Using the square's long arm, mark this on the board.

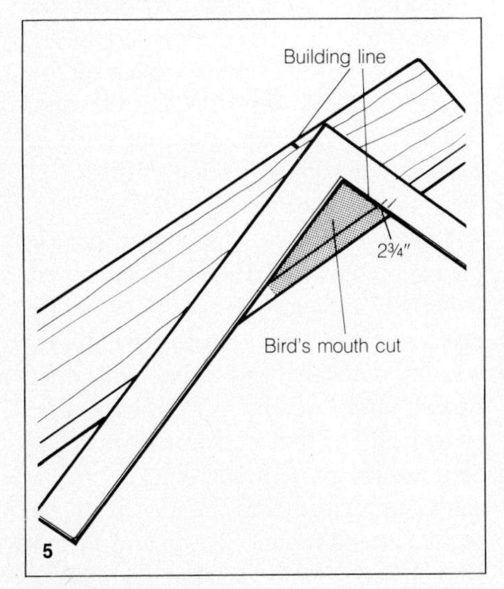

Mark where the rafter meets the wall; align the inside of the tongue with the mark to draw the building line and the bird's mouth.

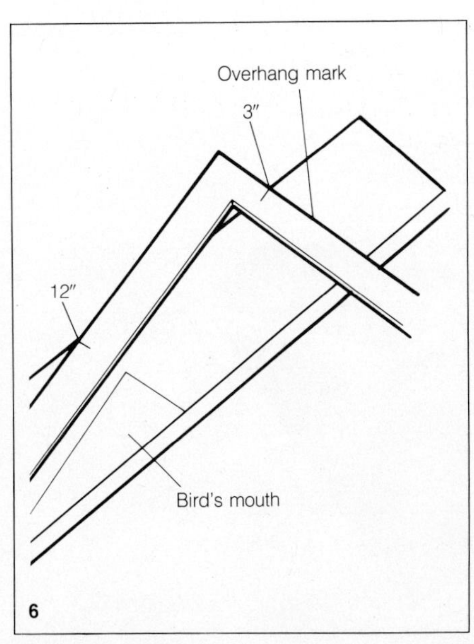

Place the 12-inch mark of the long arm on the building line. Align the 3-inch mark on the tongue with the same edge to find the length of the overhang and the cutting angle.

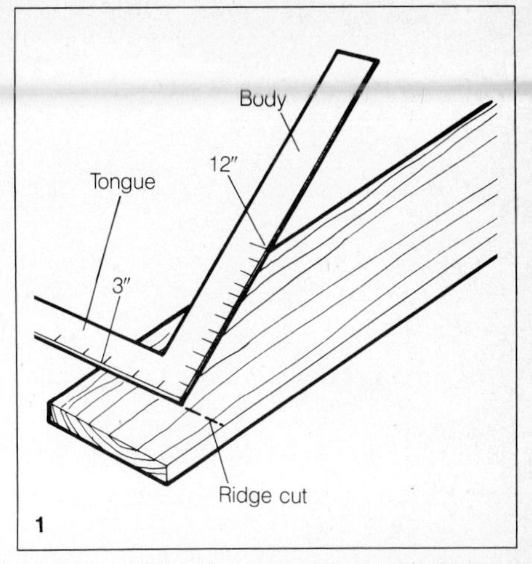

Align long arm with upper edge of board. Pivot on the 12-inch point until 3-inch mark on tongue intersects the edge of the board. Mark along tongue to show angle of ridge.

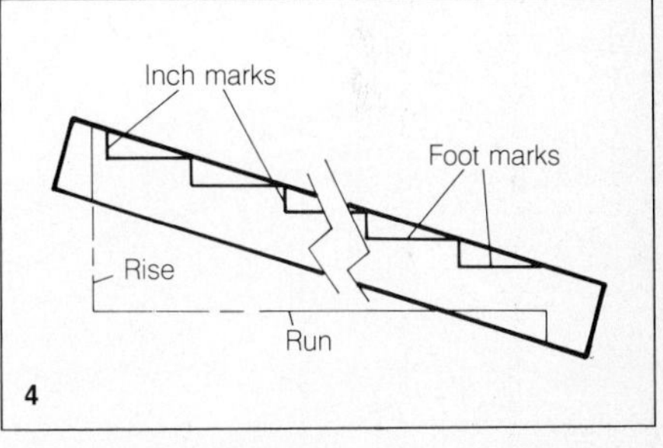

Finally, place the 3-inch mark on the tongue and the 12-inch mark on the arm, against the edge. Starting at your last mark, repeat for each foot in the run.

of the dormer and extending it to the window opening, or at least 6 inches upward. It will go on top of the first course of shingles that are below the dormer, underneath the shingles at the sides of the next course. Step flashing (see Chapter 9) is applied to the sides. Valley flashing is laid over the joint where the dormer and roof intersect. Place roofing felt over the dormer, with its ends under existing shingle. Add shingling to match the main roof. Then install the window(s) according to manufacturer's directions. Check for plumb and square and nail window frames securely to the window opening framing.

Attach building paper to the sides of the dormer, and finish with siding to match the rest of the house. Be sure that the siding goes over the step flashing at the sides, and butts tightly against the window frame. Caulk around the window and at all joints. Paint if desired.

Step twelve: insulating. Insulate inside the dormer at both walls and ceiling, using as great an R-factor as possible. Finish the interior as desired.

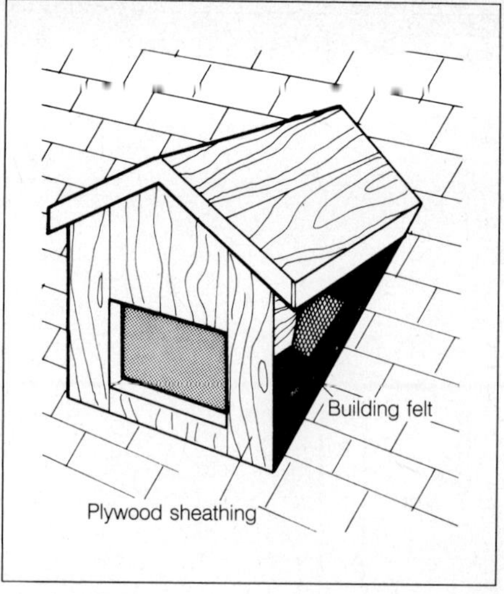

The whole dormer is sheathed in plywood, which is then covered with building felt before applying the siding or roofing.

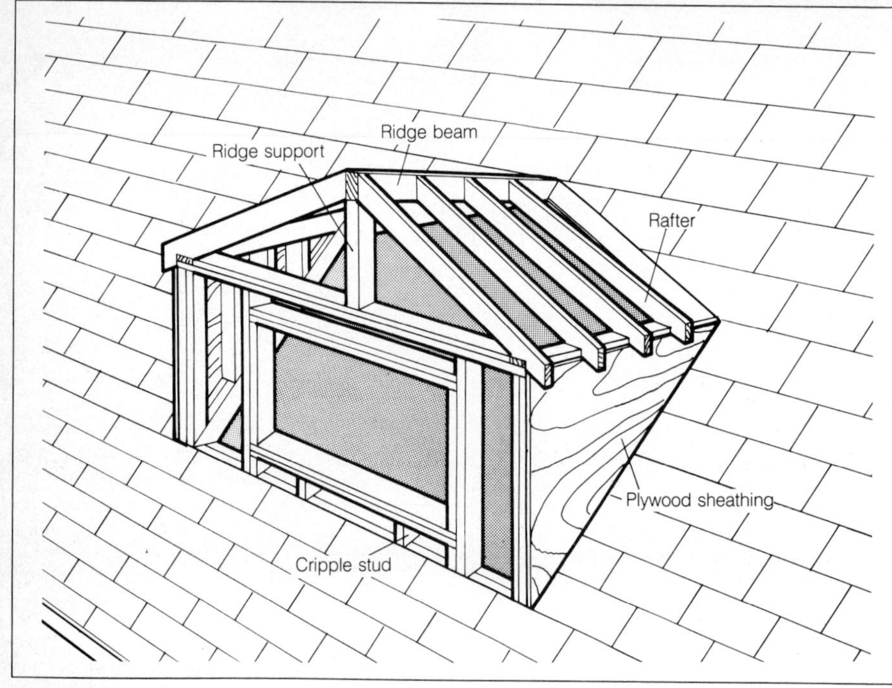

Cut the rafters to size and install. Sheath sides of the dormer with plywood. Install cripple studs below the rough window opening to provide support for the frame.

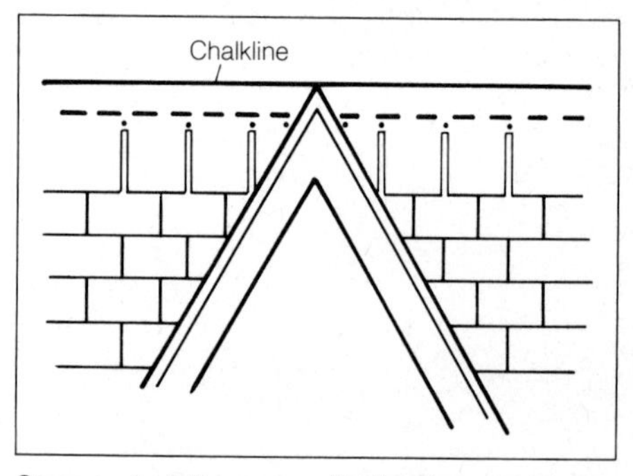

Snap a chalkline even with the dormer ridge to serve as a guideline so that shingles on main roof and the dormer roof will align.

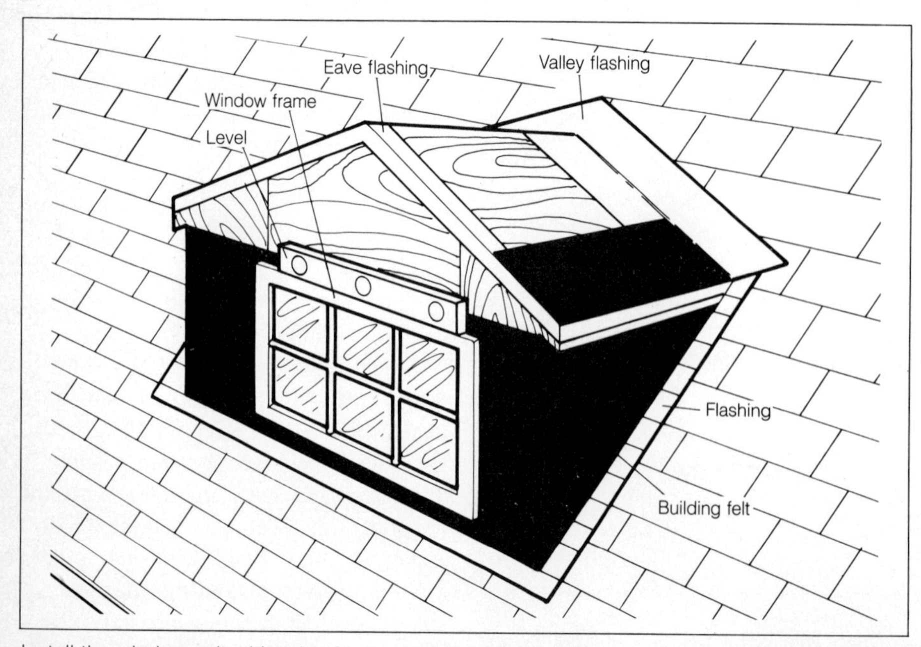

Install the window unit, shimming for level and plumb. Install eave, drip edge and valley flashing and then roof underlayment. Add fascia board; lay field and ridge shingles.

How to Build a Shed Dormer (with Help)

In basic construction, a shed dormer is probably easier to build than a gable dormer. There are no valley rafters and fewer difficult angles to measure and cut. Supporting posts are not necessary, because the new structure is completed before the existing rafters are cut off.

The problem here is that the shed dormer is bigger, and the roof must be exposed for a longer period of time. Bad weather can be disastrous if the interior is left open. When building a shed dormer plan very carefully. Have on hand all the materials necessary for completing the exterior. Pay very close attention to weather forecasts and work only when warm, sunny weather is predicted. Line up as many willing workers as you can to help the job move along quickly. If you can't comply with these requirements, the section after this one tells you how to do it all yourself, but be warned that this is difficult and time-consuming. In either case, have enough tarps or heavy plastic

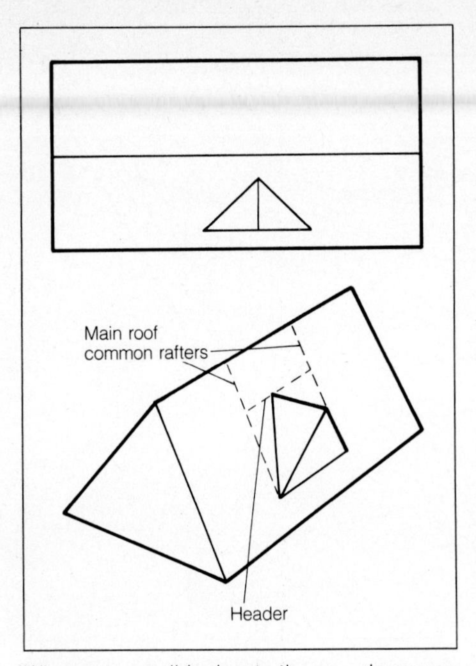

Whenever possible, locate the new dormer so that it falls evenly between common rafters. The rafters are supported by headers.

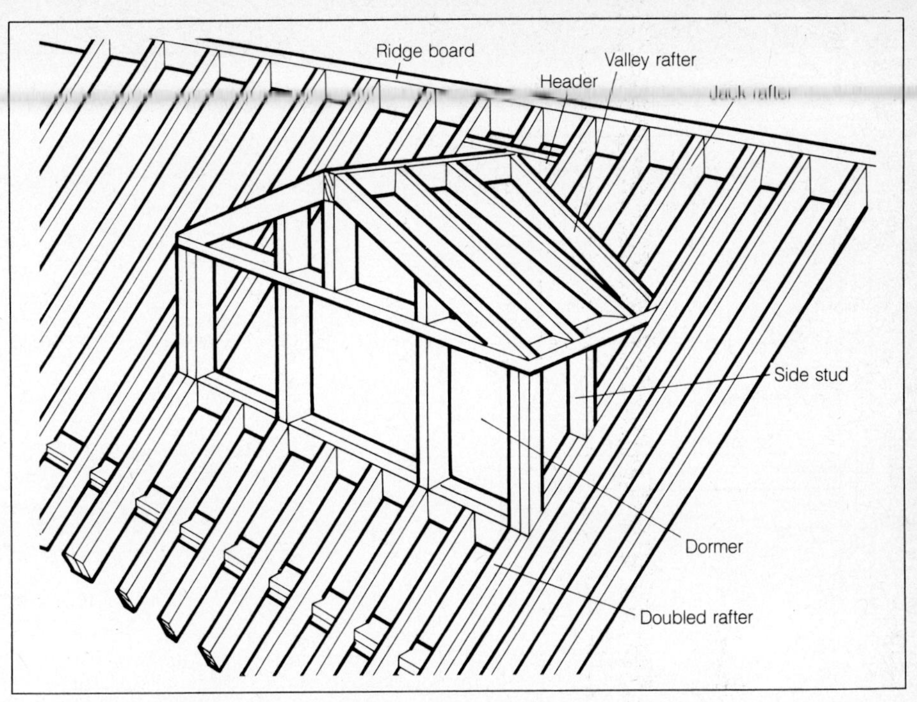

A dormer requires use of several types of rafters, as well as doubled headers.

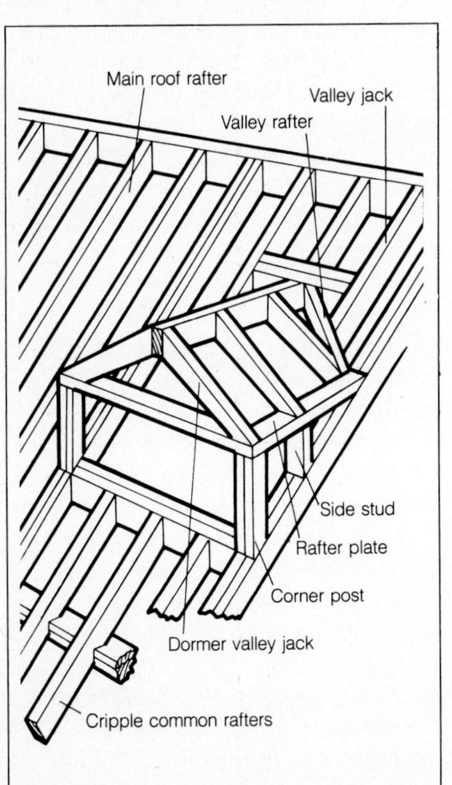

This small dormer requires only one cripple rafter set in above the header.

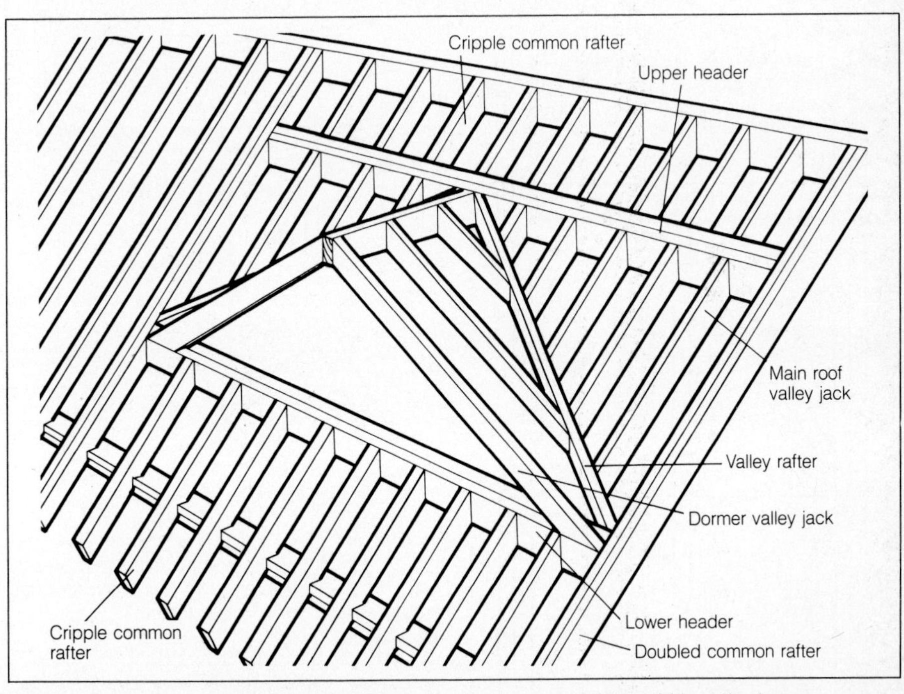

This low, wide gable dormer needs several cripples because of the width at the base.

sheets to cover the opening in case of rain.

Step one: supporting the dormer. Locate where the ends of the dormer will be, and install new doubled flanking rafters if none are there now. They will nail to the top plate of the wall and to the ridge beam. If rafters are there already, you can double them up (later) at each side.

If the dormer extends all the way to the wall of the house, the front of the dormer will rest on the existing top plate. If you are setting the dormer back, snap a chalkline on the attic floor where the new wall will rest. Cut two straight pieces of 2x4 as plates. Nail one of the 2x4s carefully along the marked line, using 12d nails to nail into the floor joists. Take the other piece and lay it next to the first one. Mark the location of the studs on both pieces at the same time. The loose piece will be used as the top plate later. If you do not need the line or bottom plate because the dormer will rest on the existing plate, just mark one piece of 2x4 with stud locations.

Step two: cutting framing members. Cut all the dormer studs to length, including the sidewalls. Measure and cut all other framing pieces, including top and bottom sidewall plates, window framing and jack studs. There will be full-height studs at each side of the opening. Place shorter studs so their ends reach to $1\frac{1}{2}$ inches under the opening. Run a flat 2x4

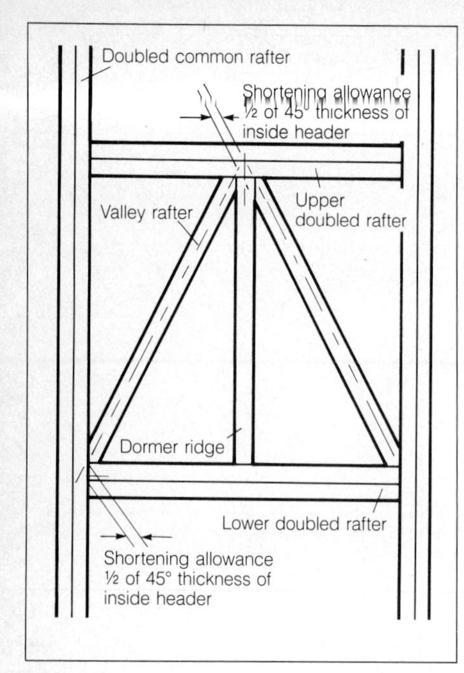

This overhead view shows how the valley rafters are cut to fit into the junction of the headers, common and ridge rafters.

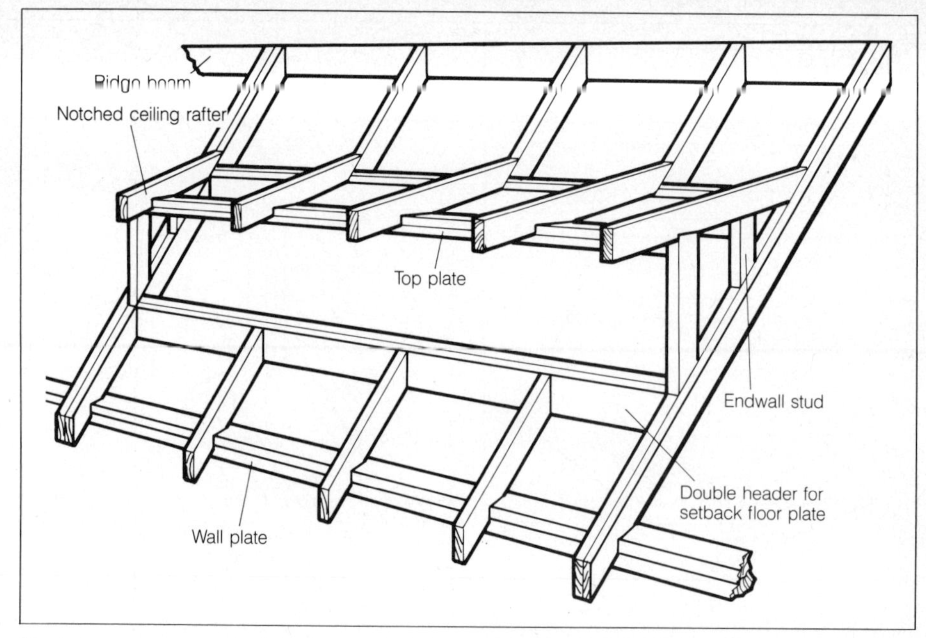

Shed dormers may be built before the hole in the roof is cut. Rafters must still be supported by doubled headers; cut gable rafters to fit common rafters and top plate.

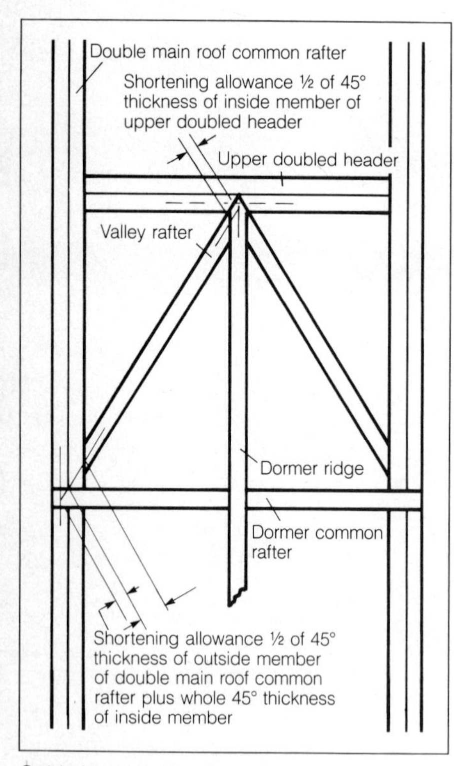

Gable beam and valley rafters may also be cut to fit junction with a slightly wider flare to the gable dormer roof.

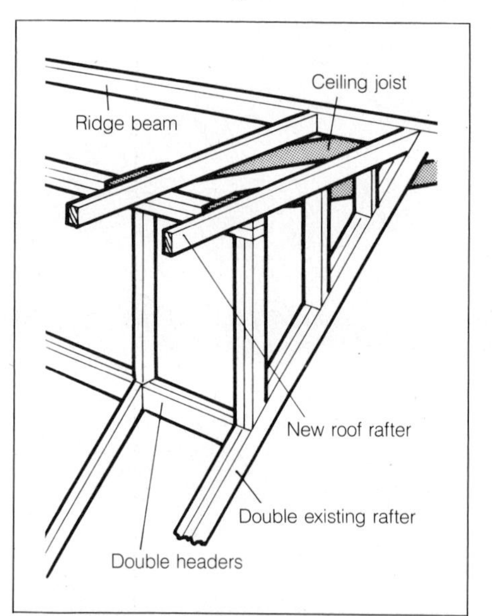

When the roofline of a shed dormer is nearly flat, the rafters do not need bird's mouth cuts. Flat rafters serve as ceiling joists.

sill between the side studs. Inside the opening, add filler studs. At the top, a doubled header is required. Cut 2x4s to fit between the top plate and the header. Number or otherwise key all the precut framing members so that everything can be located quickly when needed.

Step three: framing the opening. The opening will be flush to the double flanking rafters that run alongside the opening. However, do not begin cutting any framing until you have added the supporting framing. Have one or two helpers put the front wall together, toenailing the studs into the nailed-down bottom plate, as previously marked. Attach the top plate by nailing into the stud ends through the plate; use 12d nails. If you are not setting the dormer back, toenail to the existing top plate. Meanwhile, assemble the sidewalls separately. Set the sidewalls in place and nail them to the front wall and to the existing rafters.

Assemble the window framing for the front wall, and attach the ceiling joists. These are toenailed to the top front plate and fastened to the rafters on the other side of the roof, using 12d nails. Tack 1x2s across the tops of the joists to lock them in place.

Step four: installing the rafters. Now fasten the new rafters, starting with new doubled rafters alongside the existing flanking rafters. Nail them to the existing rafters, the ridge board, the top plate and the joists. Use 16d nails. Nail the intermediate rafters to the joists, the ridge beam and the top plate, but not to the existing rafters.

Step five: opening the roof. Now you can open up the roof. Start the cut next to the flanking rafters. Use a keyhole saw to drill a starter hole and then finish with a crosscut saw or reciprocating saw. For plywood sheathing, make horizontal cuts in the plywood sheathing to help you pry the panels out. For board sheathing, knock the lumber out from the inside.

Step six: cutting the rafters. When the dormer is flush with the front of the house, the framing is now complete, and old rafters can be removed. If your dormer is set back, first nail the existing rafters to the front studs. Then cut the existing rafters flush with the inside of the studs. Then pry or cut the other ends from the ridge board.

Step seven: adding the sheathing. To nail sheathing across the rafters, start at the edge of the roof. The ends of the rafters should overlap by the thickness of the fascia board and the wall sheathing.

This is the minimum overlap; the exact length depends on the roof overhang. Next, sheath the dormer walls. Apply building felt to the sheathing. Position the windows; check for level. Place flashing for the dormer (see Chapter 9).

Building a Shed Dormer (By Yourself)

When there is work to be done, many of us suddenly become friendless. If this seems likely in your case, it is possible to complete the job by yourself, but it isn't easy. You will have to install the dormer while keeping the old roof in place. Then, over a period of time, you can remove the roof from the inside.

Step one: preparing the framing. You will be cutting small holes through the roof for each vertical framing member. This is easier when the front wall of the dormer will be above the existing wall. All you have to do in this case is remove the decking just up to the top wall plate, and lay the new dormer front wall on top of that. If the dormer is a set-back arrangement, lay the bottom plate as for the shed dormer described above. Careful planning and measuring is a must when you work this way.

Cut holes in the roof through which 2x4 studs can be run. Toenail the end and intermediate studs to the floor plate or top of the wall plate. If the new wall is exactly above the old one, do not nail the studs to the old rafters; they will be removed later.

Step two: finishing the framing. Lay the top plate across the studs and nail into the ends of the studs. Install new rafters on the flanking rafters, as for the shed dormer discussed above. Frame in the front opening. Add the new decking and roofing on top of the new rafters.

Step three: placing the endwalls. Removing only as much of the roof as you have to, nail on the top plates for the end walls. Attach them to the top plate of the front wall and to the rafters that were attached to the flanking. Finish the front wall window framing. Install windows and finish the exterior of the wall. The dormer should now be weathertight and the removal of the sheathing and old rafters can proceed from inside at a more leisurely pace.

Step four: removing the old roof. Now the old roof can be removed. This will have to be done piecemeal, a bit at a time, because of all the new framing. Do not just throw the old pieces out the new window—arrange to lower them safely or cart them through the house to the outside.

Protecting the roof opening. Up to this point, there should have been no large sections of the old roof exposed to the weather. When you do take a break, however, protect the small holes in the roof. Place scrap pieces of plywood, building paper or heavy plastic over all areas where holes have been cut in the existing roof.

EAVE EXTENSIONS

A roof overhang serves several functions. It gives a more pleasing appearance than eaves that are chopped off at the exterior wall. It protects the siding and foundation plantings where there are gutters, and it helps prevent ice-dam buildups.

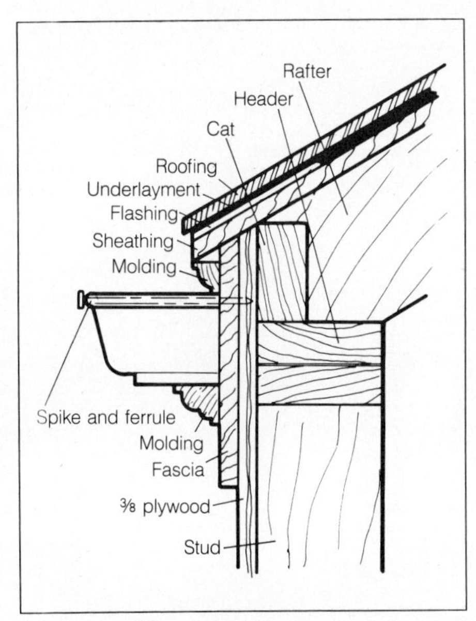

A short overhang reaches beyond the building line to direct rainwater. Overhangs of insufficient length can lead to ice dams.

Ice dams form when the warmed snow on the roof melts and the water flows down into the cornice. There it again freezes and blocks the further flow of melted snow, which can back up into the eaves and down into the walls and into the house.

If your roof has little or no overhang, it is a good idea to extend the eaves before reroofing. When the eaves are high enough, they can often be extended far enough to provide a roof for an extra room, patio or deck. This type of extension will involve structural changes outside the scope of this book.

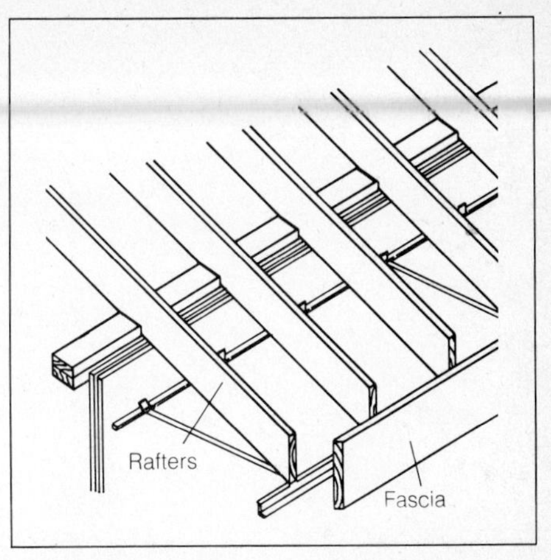

A longer overhang provides protection to the siding and gives shade during the summer. Rafter ends are covered by a fascia board.

Extending the Eaves by Yourself

You can extend the eaves of your roof by a foot or two without cutting into the roof structure, avoiding extensive alterations of the rafters.

Study the existing eaves and determine the angle needed to add rafter extensions. Using pieces of 2x6 lumber, build out from the existing eaves to the desired length. Then add new fascia boards. Build up the decking to the level of the existing deck. Add new shingles along with the rest of the roof; install new gutters and leaders as required.

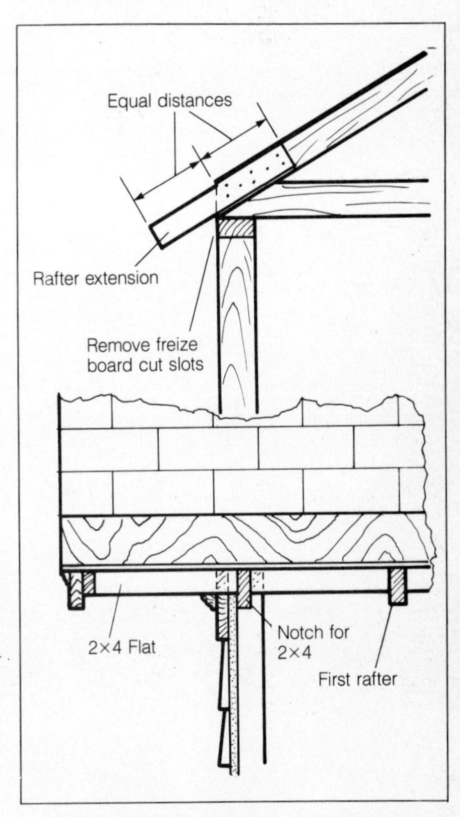

Lumber for eave extensions of existing construction must be twice as long as the overhang itself. Then add decking for roofing.

A dormer adds considerable usable space to the second floor of a small bungalow. This dormer is faced with stucco, a durable surface that offers good insulation.

The attic space of this hip roof is enlarged by this dormer that also admits natural light to the attic.

This home has a roof extension over an entry and a dormer in the second floor level. Rooflines have been repeated for a unity of style.

This is an eyebrow window, a style of small dormer to provide light to a little used area under the roof.

A large dormer provides space in a second floor expansion. Gable dormers are usually smaller and set back farther from roof edge.

A shed dormer fits into the planes of a gambrel roof. The shed roof is a continuation of the plane of the upper roof. The dormer substantially increases the second floor space.

Rather than adding dormers to provide light and ventilation to this hipped roof attic, the owners installed a series of skylights.

On a relatively flat roof, a skylight may angle the glass toward the sun for more light.

A skylight provides plenty of light while retaining maximum privacy for this attic bath.

Longer extensions. For longer eave or rake extensions, cut into the frieze or fascia boards along the existing rafters so that the extension boards can be inserted. Use a saber or keyhole saw and make the holes the same nominal size as the rafters (for 2x6 rafters, cut out holes 2 inches x 6 inches).

Extension rafters should be as long inside as out. If, for example, you are making a three-foot extension, cut the rafters six feet long. Insert the extension rafters into the holes previously cut, and nail to the existing rafters from inside the attic, using 12d common nails every 12 inches. If you do not have access to the rafters from the attic, you will have to cut holes in the deck and nail from above (or forget the whole idea).

Use plywood or sheathing boards to deck the extension. If you are not removing the present shingles, you can make the deck the same thickness as the present deck plus shingles (assuming the whole roof will be reshingled). If the old roofing is (or will be) removed, the deck material should be the same thickness as the present deck. When the extension deck is the same level as the current surface, apply roofing as previously directed.

ADDING A SKYLIGHT

No-cost illumination is understandably welcome in an age of constantly rising electricity costs. A skylight offers a practical and attractive means of flooding a room with natural light without loss of privacy.

On the negative side, a skylight can be a security problem; it can leak; it may cause great loss of heat in winter and promote heat buildup inside the room in summer. It can be difficult to clean and maintain. Local security and building codes may restrict their use. Be sure you know and understand the local regulations governing size, placement, material, safety, and shape before you make a final decision.

Selection

Before the advent of glazing plastics, skylights consisted of flat panels of tempered sheet glass, either clear or frosted, often reinforced with wire mesh, and supported in metal or wood frames. Today, skylights made of plastic in dome and bubble shapes come from the factory complete with metal frame, anchoring

Most skylights come complete with all framing and flashing and ready to install.

flanges, and watertight seals. Installation does require you to cut into your ceiling and/or roof, so you may wish to leave this to an experienced carpenter if you have not previously done this type of work.

Ready-to-install plastic domes can be placed flat over a lightwell, or situated on a sloping roof. In either case, domes are easier to keep clean than flat sheets of glazing because rainfall tends to wash away accumulations of dirt and grime. Choose skylights with double glazing to cut down on condensation.

If local codes permit, and you feel there would be no security problem, select a skylight that can be vented. Some varieties of plastic domes are made to be opened from the inside. This design is ideal for proper ventilation, and for control of humidity and condensation. Flat skylights can be made to open and close by a system of pulleys, pivots, and/or sliding mechanisms.

Cutting the Roof Opening

Carefully read all instructions, which will call out the size of the opening necessary for that particular model. If the opening goes directly through the roof, as in a ceiling which is also the underside of the rafters, both the ceiling and roof openings are the same size. If the skylight will require a shaft to direct the light from the skylight in the roof to an opening in ceiling—where there is an attic—then the ceiling opening will be a different size. The chart lists the required openings for the various sizes of skylights, in both the roofs and in the ceilings.

Finding the slope. To use the accompanying chart, find the slope of your roof. Use a level to measure one foot along the roof, horizontally, from a point where the level intersects the sloping roof. Mark the roof at the intersection and the one-foot point. At the latter mark, use the level (or a ruler) to find the vertical distance between the one-foot mark and the point on the roof slope directly above it. You will have created a right triangle which intersects the roof slope at two points, and which gives you the rise-to-run ratio.

Measuring for ceiling cut. Try to avoid cutting into joists or rafters, although it is likely you will have to cut through at least one. To minimize the problem, and to avoid having to cut two joists or rafters, determine roughly where you want to locate the skylight and then bore a ⅛ inch hole through the ceiling near the center of the skylight location. Push a length of wire through the hole, then go upstairs and check where the wire is in relation to the floor joists. (If the ceiling is against the underside of the roof rafters, you will have to use the time-honored method of tapping on the plasterboard to locate the rafters.) Measure from the projecting wire, down against the plasterboard, to the joists on either side. Then go downstairs and repeat the measurement on the ceiling. If it is possible to keep the skylight opening between joists by moving it an inch or so either way, then do so; you will save time and labor.

Cutting the ceiling. Mark the opening on the ceiling and cut along the lines with a keyhole saw or handsaw. Hold up the plasterboard as you cut the last side, to keep the saw from binding as the board starts to tilt down. Wear safety glasses, and mask, or a face shield, to keep plaster dust out of your eyes and lungs. Even if you cut away from yourself so that the dust does not fall on your face during cutting, you can expect the plasterboard to drop afterward and create a shower of dust and debris.

Cutting The Roof. At this point, measure up from the opening you have cut, using a plumbed straightedge, at each corner; mark the underside of the roof. Drive a nail through the roof at each marked corner, then go up onto the roof. Leaving the nails in place, remove the roof shingles inside and about 6 inches to a foot around the opening marked by the nails. Now snap a chalkline from nail to nail, or mark along a straightedge from nail to nail.

Remove the nails and cut along the lines. A portable electric saw will work well for this job, but stop just short of the corners and finish the cuts with a handsaw to avoid cutting beyond the corners. Even the largest of portable electric saws will not cut completely through a rafter; it will have to be cut loose with a handsaw. Make sure, of course, that no one is standing under the opening to be hit by the falling lumber. Check also that there

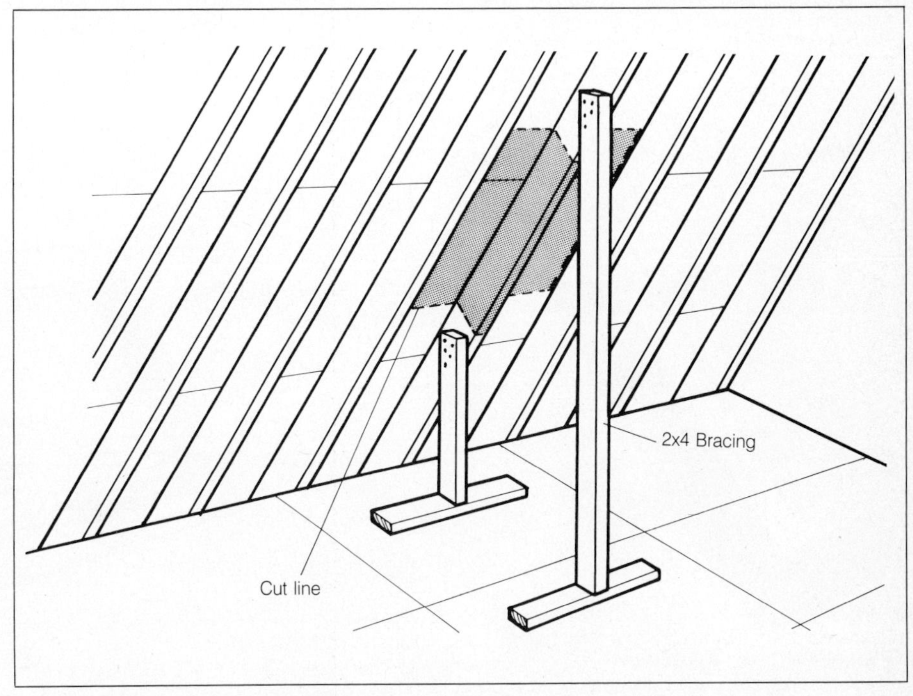

Before cutting through the roof, and particularly a rafter, prop the rafter well. Use two 2x4 braces nailed to the rafter above and below the proposed cut.

is no furniture or any other item below which can be damaged. One safety procedure would be to nail a stout piece of rope or light chain to the roof at one side and to the section of roofing being cut loose. When the final cut has been made the piece will fall in, but will be held by the chain or rope. A helper then could lower the piece after prying up the nail on the roof, and there would be no danger. To work safely, think several steps ahead and realize the possible consequences of each operation.

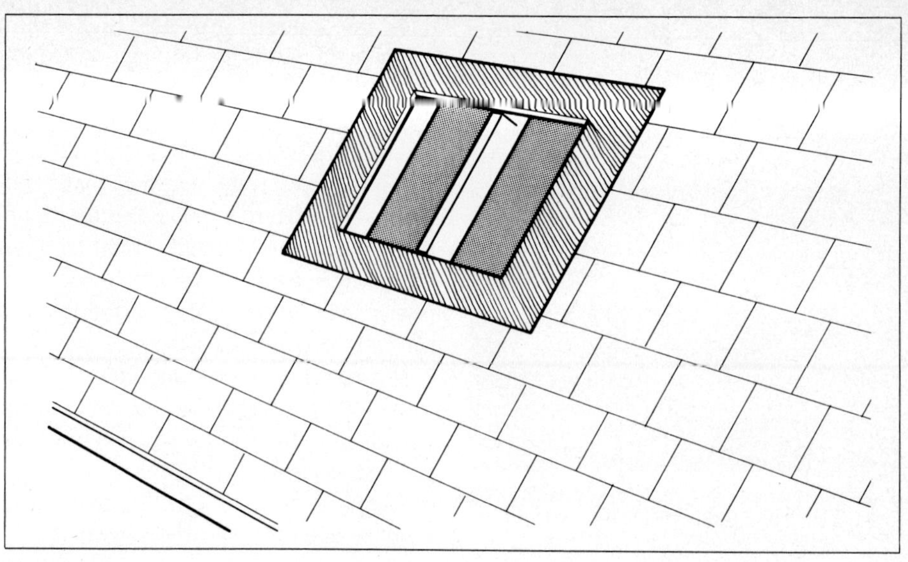

Shingles are removed and an opening cut in the sheathing. Rafter is exposed for cutting.

Framing The Opening

Frame in the opening you have created, using lumber of the same size as the rafters or joists. Double up the framing at right angles to the joists. This may require cutting another 1½ inches from the ends of the joist or rafter that was cut. The additional ½ inches will allow for 12-inch joists at each end. If a light shaft is required down through an attic, build it from the 2-inch framing down. It probably (almost certainly) will be necessary to cut a ceiling joist and box in the opening as in the drawing.

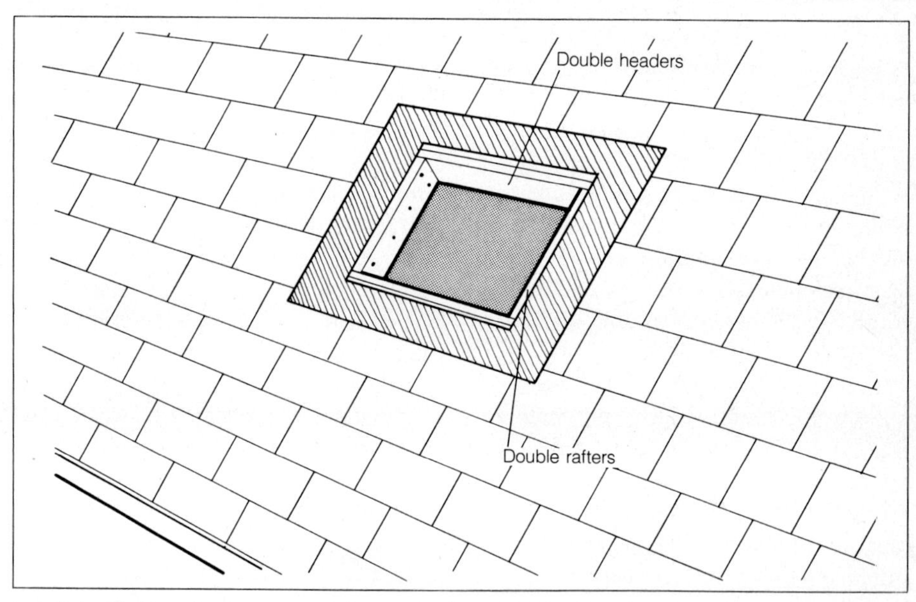

Install doubled headers across opening; next add a doubling board to each of the rafters, at each side of opening.

Positioning The Skylight

Now spread roofing mastic or compound around the opening on the roof. Keep it away from the opening so none will run down inside. Position the skylight over the opening so its corners are aligned with the corners of the opening. It will be hard

Apply roofing compound to seal cracks between new framing and roof sheathing to prevent joint leaks after the unit is in place.

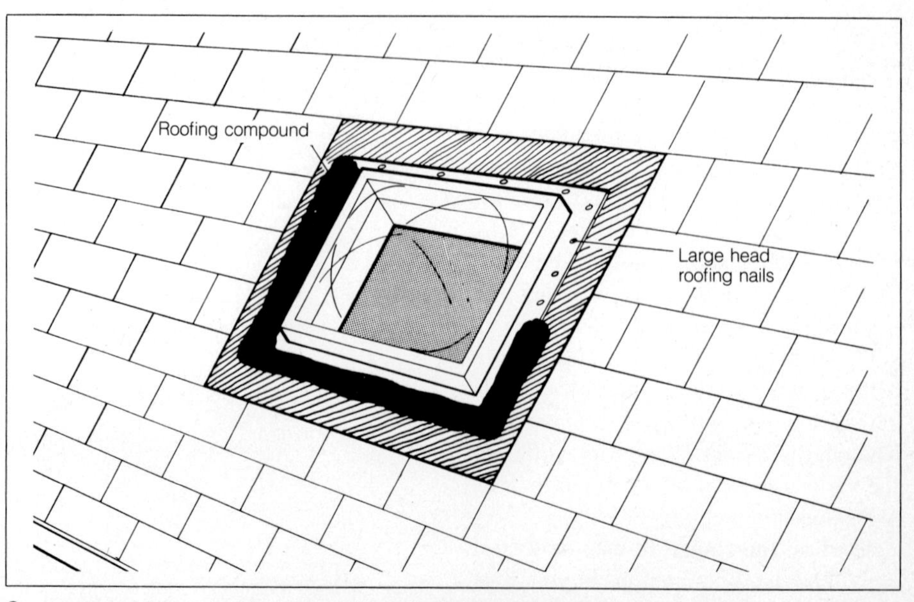

Cover edge of flange and fasteners of installed skylight unit with roofing cement.

to keep your knees out of the mastic, so ask a friend to help with this step.

Once the skylight is in position, press it firmly into the mastic. Nail through the flange into the roof. Use large-headed shingle nails that will not easily drive through the aluminum flange.

Finishing Off The Opening

Replace the shingles and apply roofing compound between the shingles and the curb of the skylight. Cover the flange completely with mastic before applying the shingles, and cover all of the nail heads with dabs of mastic. The last step can be the application of a bead of roofing compound between the edges of the shingles and the skylight from a cartridge of the compound, using a caulking gun. This is quick and easy, and is less messy than scooping out the black liquid tar from a container.

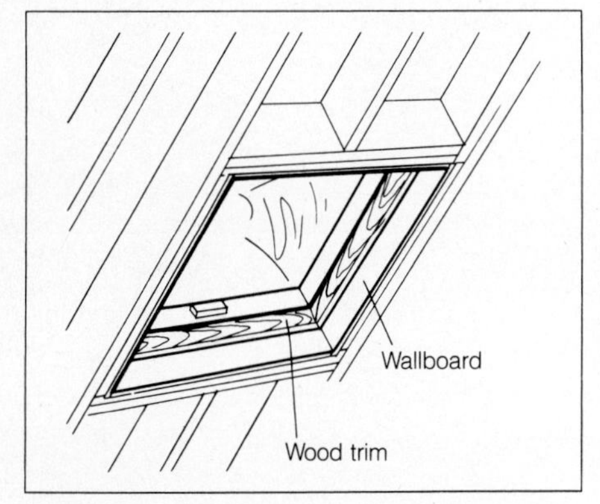

Finish opening interior with appropriate materials, such as wallboard and wood trim.

If the skylight has been installed in a roof with a ceiling on the underside, finish off the inside of the opening with plasterboard, paneling or other material.

Finish off the plasterboard with paper tape and joint compound. For outside corners, use metal corners, especially if a hinged drop-down window is used at the bottom of the light shaft. The drop-down window gives the ceiling a more finished look, and avoids the sometimes objectionable appearance of a "hole in the ceiling" which a light shaft can produce.

Once your plasterboard has been taped, and the two coats of compound applied, dried, and sanded smooth, then brush on a sealer. Paint the taped joints and edges.

Creating a Light Shaft

If there is a space between the roof and ceiling below, make and install plywood box to serve as a light shaft.

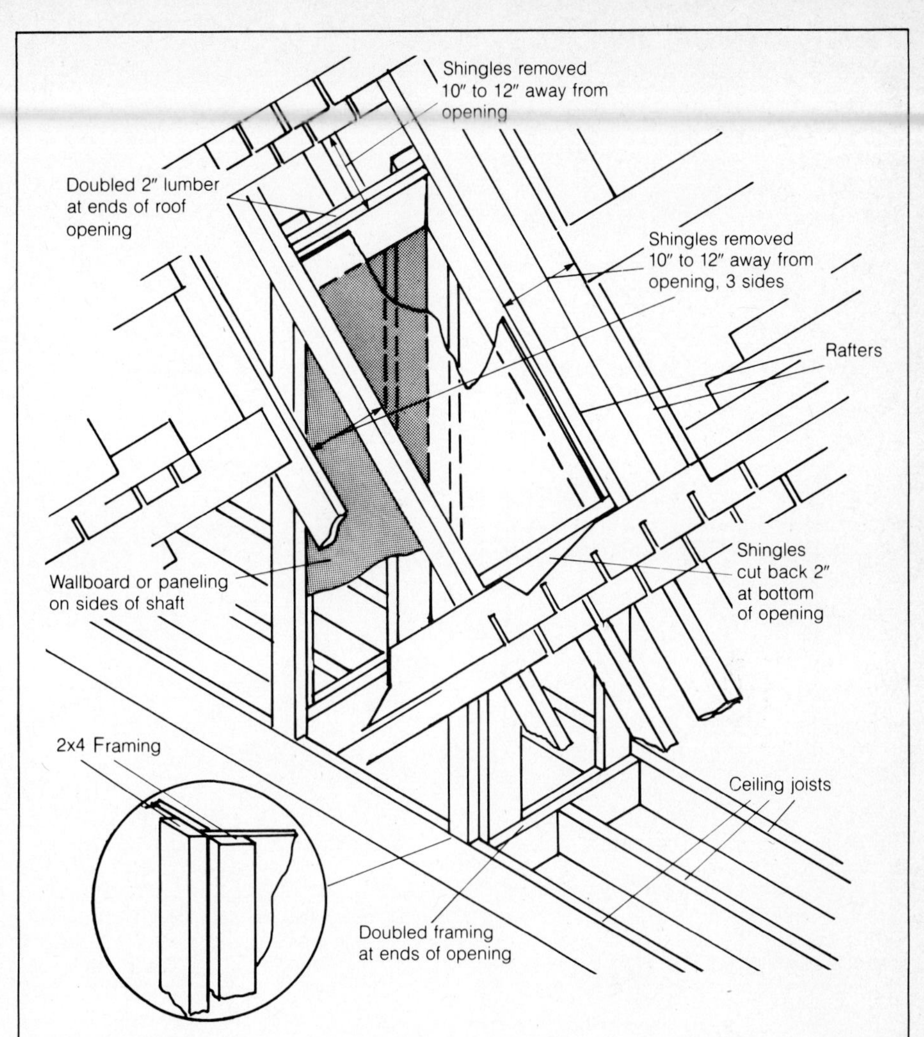

For an attic, build a light shaft between roof and ceiling. The ceiling opening is shorter (in line with roof slope) than the roof opening. Drop a plumb line from the corners of the roof opening to locate the ceiling opening, or project a straightedge board up from each ceiling-opening corner to find the roof opening. Plumb the straightedges 2 ways with a level to align the 2 openings. Frame the light shaft as for a wall, using 2x4 studs. In unheated attics, install insulation on outside of the light shaft.

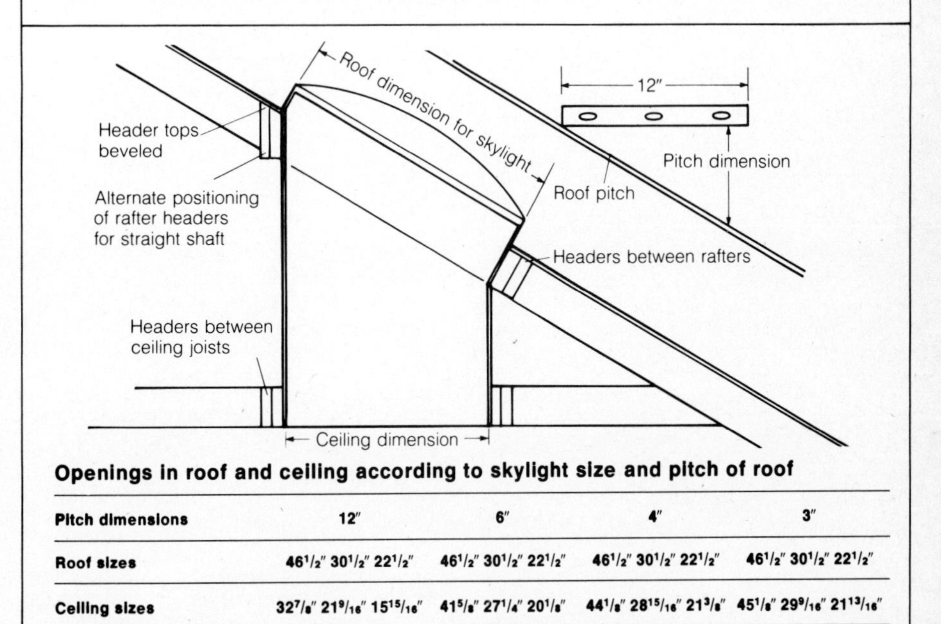

Openings in roof and ceiling according to skylight size and pitch of roof

Pitch dimensions	12"	6"	4"	3"
Roof sizes	46 1/2" 30 1/2" 22 1/2"	46 1/2" 30 1/2" 22 1/2"	46 1/2" 30 1/2" 22 1/2"	46 1/2" 30 1/2" 22 1/2"
Ceiling sizes	32 7/8" 21 9/16" 15 15/16"	41 5/8" 27 1/4" 20 1/8"	44 1/8" 28 15/16" 21 3/8"	45 1/8" 29 9/16" 21 13/16"

Chart gives the sizes of openings in roof and ceiling according to the skylight size and the pitch of the roof.

CONSTRUCTING A DECK ON A PORCH OR GARAGE

In some urban and some older surburban areas, original building codes specified a minimum lot size that was so small that many homeowners did not have sufficient yard space to enjoy many outdoor, leisure activities. If the existing home on this type of lot has an attached garage or an enclosed porch, it is often possible to use this roof area to provide a surface for an attractive, useful second-floor deck.

Skylight illumination can be directed to a lower floor through a lightshaft. This plywood box constructed to fit against skylight frame and an opening in the floor. The inside is painted white for reflectivity.

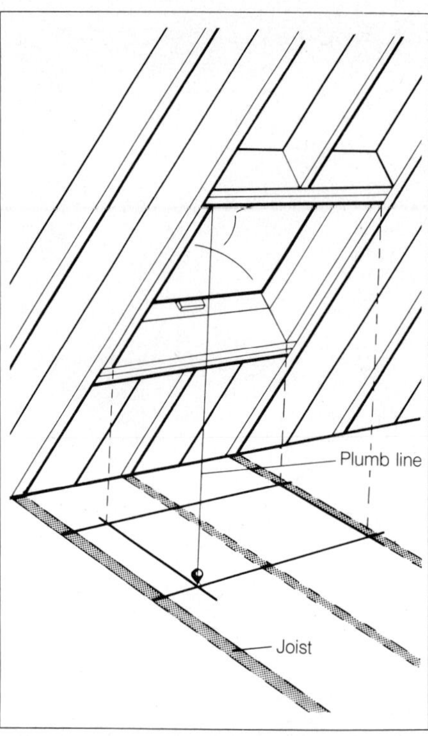

To mark position of the opening to a lower floor, attach a plumbline to the corners of the skylight framing and mark the floor. These marks will serve as guides for cutting the correctly sized opening for the shaft.

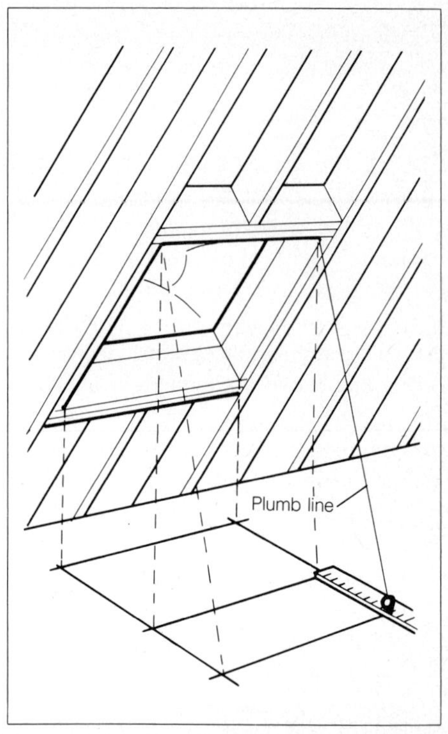

The area of the lightshaft may be larger than the opening of the skylight. The area may be enlarged by laying a straightedge along the opening line parallel with the rafters and moving plumbline to mark corner.

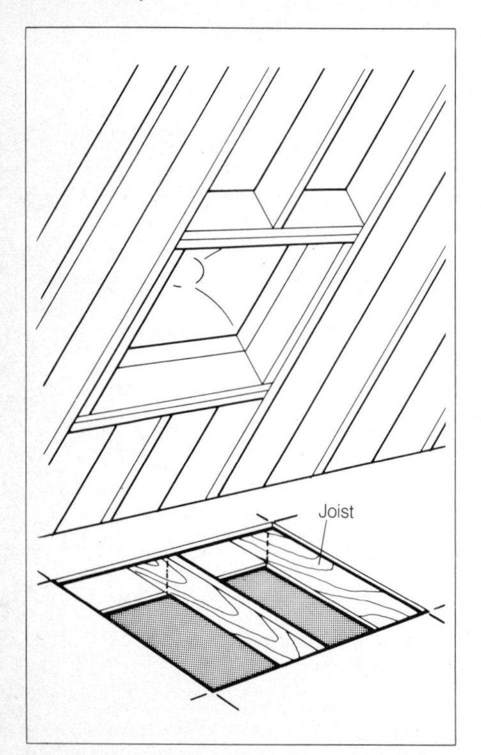

To locate correct position of the doubled joist, drop a plumbline from the skylight frame and align a board below the plumb bob. Use a straightedge to mark the line on which the doubled joist is to align.

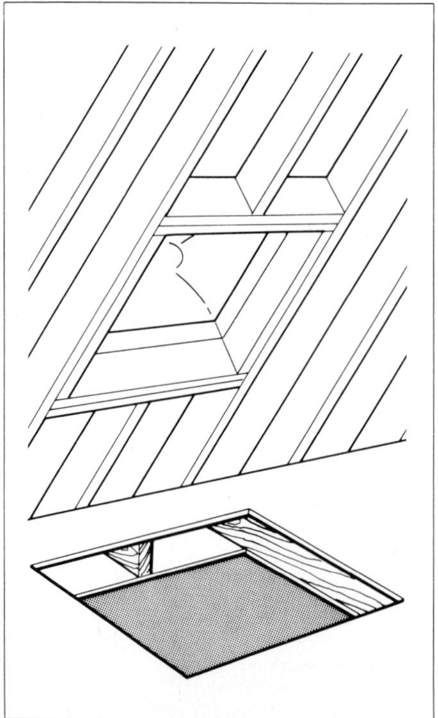

Prop the ceiling from below and cut through joist to be removed. Opening is larger than the lightshaft to allow for doubled headers and doubled rafters needed to maintain the integrity of the ceiling.

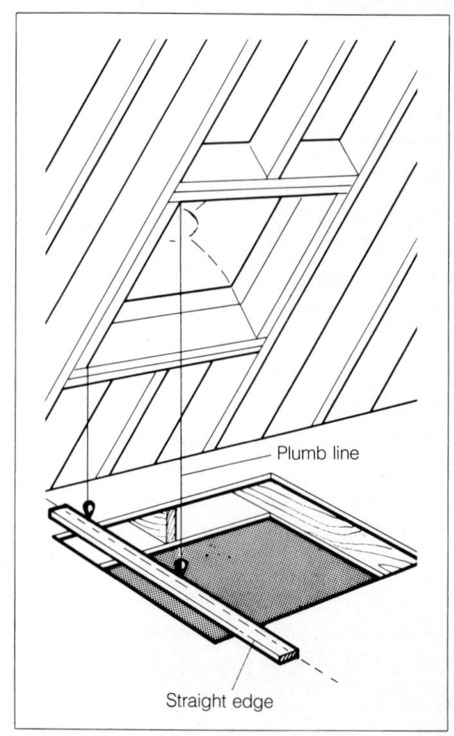

Cut away any flooring that may be in place over the opening. Drive a nail through ceiling material of room below to locate corners of opening. Cut away ceiling material between marks. Place cutting marks on joist.

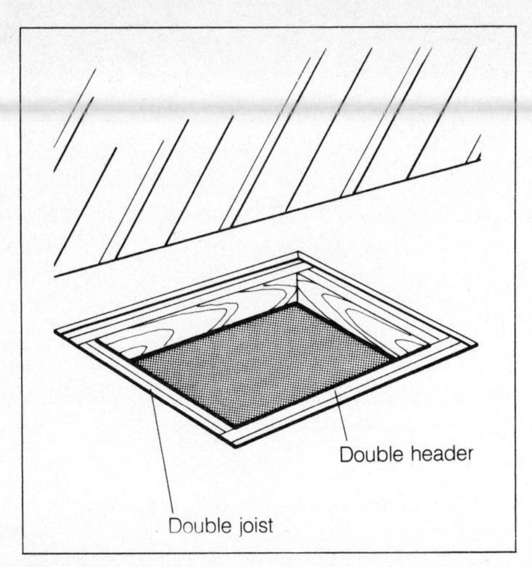

Frame the lightshaft opening with doubled headers across cut joist ends. Next double the joists on either side of the opening.

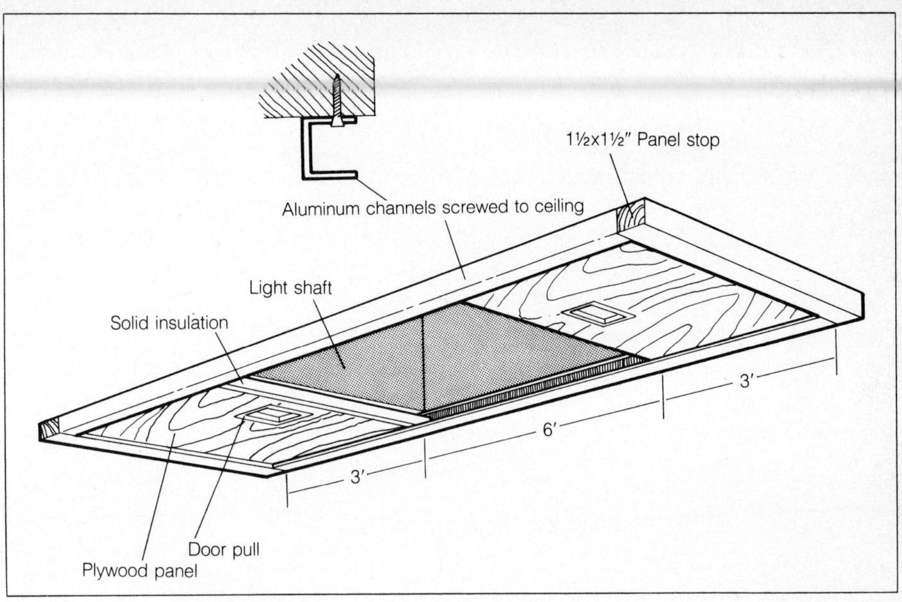

To darken a room or to prevent heat loss during the winter, install a sliding track and plywood shutters over the opening.

Install a 2x3 sole plate. To mark angles, set boards next to plate and rafters and scribe-lines. Install on 16-inch centers.

A rooftop deck can be added to an attached garage or to a detached garage on a sloping lot. Be sure to plan for access to the deck.

An attached garage is a logical place for a second story deck. The railing provides an attractive visual finish to the otherwise plain, flat roof.

If the basic roof structure is sufficiently strong and the pitch of the roof is not too steep, you can construct a deck surface for the roof. This should allow you an outdoor area for sunbathing, relaxing or even cooking and entertaining. Some older homes may have existing access to the roof. Second floor doors to an "airing porch" used to be common when maids and cleaning ladies would step outside to shake their mops and dustrags. Because the airing porch was not intended for any recreational use, the surface was usually just roll roofing.

Before pursuing any building plans for adding a deck to your porch or garage roof, check with your local building department to find out what codes restrictions, if any, govern use of these areas. You may find that you will have to add special safety railings or an escape ladder. You may also find that such use is illegal if the deck is visible from the street.

Deck Structural Requirements

The first thing you must find out is the dimension size of the lumber and the structural spacing of the rafters in the roof of the garage or porch you plan to use for your deck. This is a simple process in a garage; just go inside and look. If you plan to put your deck on the roof of an enclosed porch and this porch has a finished ceiling and enclosed overhang, you will have some difficulty in finding the size and spacing of the rafters. Do not depend on the porch roof having the same size rafters and spacing as the main roof structure, usually visible in your attic. If the porch roof was not expected to carry more than a light load, it may have been constructed of lighter framing than the main roof. You may have to pry off the fascia board and possibly part of the soffit covering to see the rafters. If the rafters are 2x6s on 16-inch centers and the porch is nine feet or less in width, this framing should be able to support the new deck and the people who will use it. If the span is greater than nine feet, but less than eleven feet-eight inches, you cannot build a deck unless the rafters are at least 2x8s. For a span greater than this, the roof will have to be even stronger.

If you discover the rafters are centered at more than sixteen inches or are of lighter framing lumber than specified above, you do not have a strong enough structure to hold the weight of the deck plus those using it.

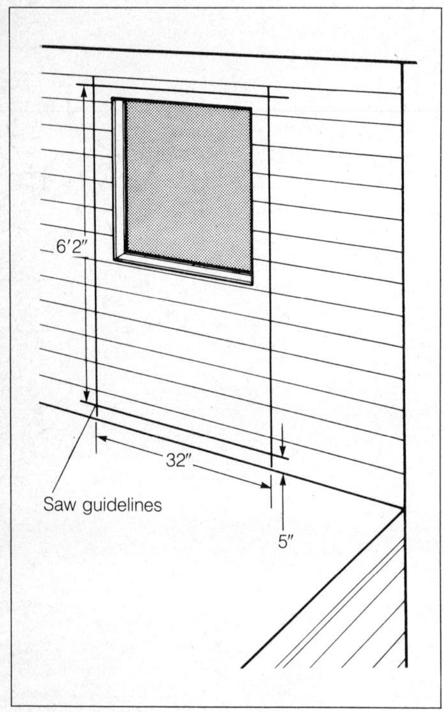

Cut basic opening for your door, leaving a sill high enough to be above finished level of the deck floor.

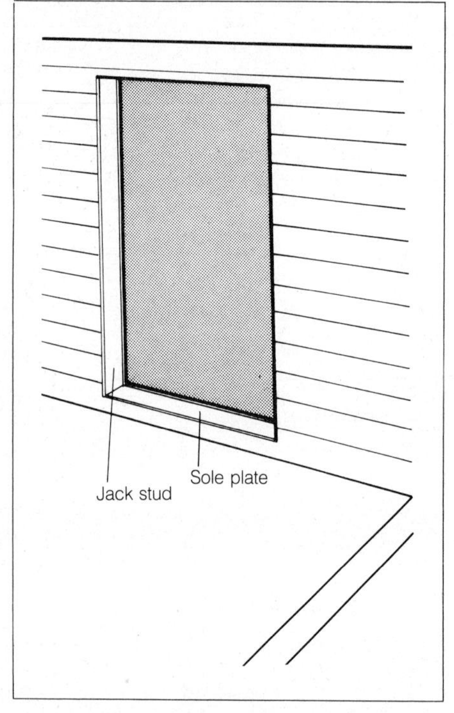

Frame up the rough opening to the specifications of the door unit you buy. Install jack stud and new sole plate at threshold.

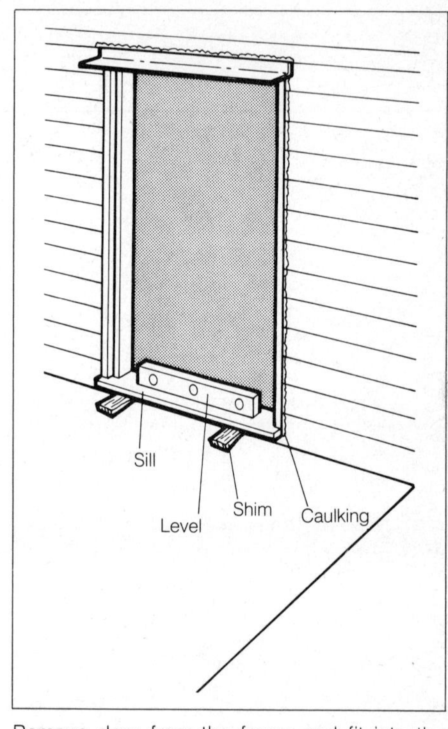

Remove door from the frame and fit into the rough opening. Shim to level and plumb. Caulk around the frame and install drip cap.

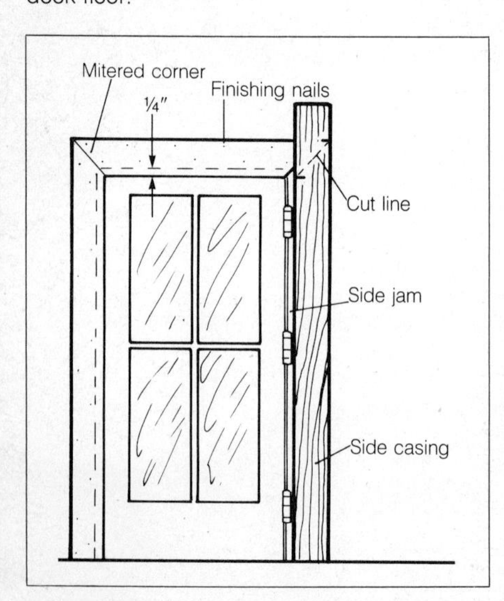

Install inside finish framing with molding appropriate to your home style. Miter cut corner joints for the best looking finish.

However, if the building code does allow a second-story deck and the rafter construction and the roof pitch meet the structural requirements, you have an ideal location for a simple, attractive and relatively economical addition to your home. The second floor location is often pleasing because it catches above-the-ground breezes, offers a different vista and, normally, is freer of insects than a ground-level deck.

Compensating for roof pitch. The basic construction of a deck becomes more complicated if the roof area is large, because more complicated support framing is required for the deck itself. For example, a deck over an attached, two-car garage often must be designed by an architect because the design must compensate for the pitch (slope) of the roof. Even the smallest slope, ¼ inch per foot, would give a difference in level of five inches from one side of a two-car garage roof to the other. For a large deck with a slope of ½ inch per foot or more, this difference would have to be compensated for by the structure of the base framing, or the deck would have an uncomfortable cant to one side.

Condition of the existing roof. Because you will be placing the deck framing directly on the roofing surface, be sure to check the condition of the roof before doing any further work. It is pointless to build your deck, even one that can be broken down for access to the roofing surface, if you are going to have to take the structure apart to make roof surface repairs within a year or two.

Check the surface for wear, blisters, rips or tears. Make any necessary repairs. If you have any serious doubts about the condition of the roofing, now is the time to do a reroofing job. Caulk the line between the roof and siding, sealing any flashing or installing new flashing if needed.

Planning the Access Door

If you are modifying or improving an airing porch, then you already have a door. However, if you are merely laying a deck on top of a garage or enclosed porch roof, you will undoubtedly have to create a doorway.

Poor locations. The location of your door is partly a matter of choice, but it must be located where it will not cause any structural problems. Bear in mind also that if the door is cut in a bedroom wall, you may find yourself disturbing a family member whenever you do late evening entertaining. An access problem like this may result in lack of use of the deck.

Desirable locations. The ideal place for the door is at the end of an upstairs hallway. However, few homes these days are constructed with long, upstairs hallways. The next best location would be in a relatively unused guest room. If neither location is possible, consider shifting room assignments so that the person whose bedroom contains the access door is unlikely to be disturbed.

Cutting the Door Opening

If a window overlooks the roof, use this as the location for the door, installing a

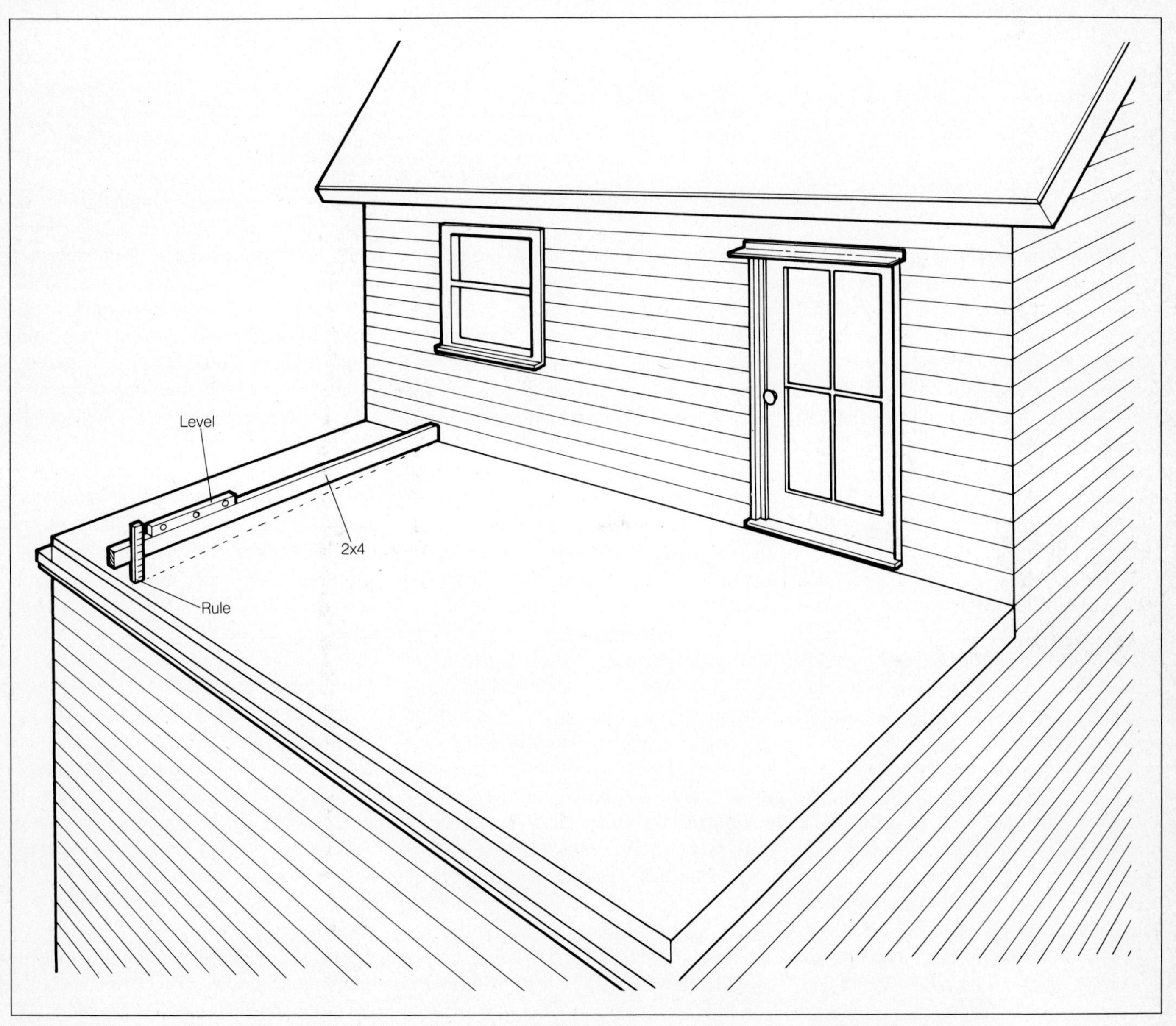

Your roof will have a slope and your deck should be level. Measure slope by lifting a 2x4 to level and noting height.

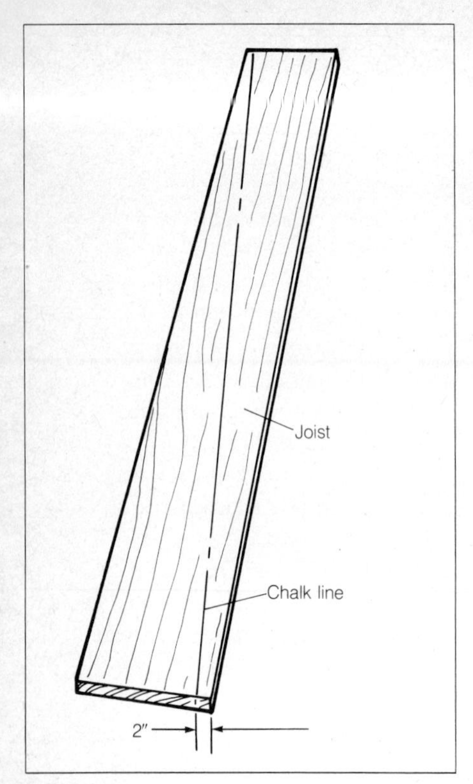

Mark the foundation boards to allow even level deck. A 2-inch height should allow for door to clear finished deck level.

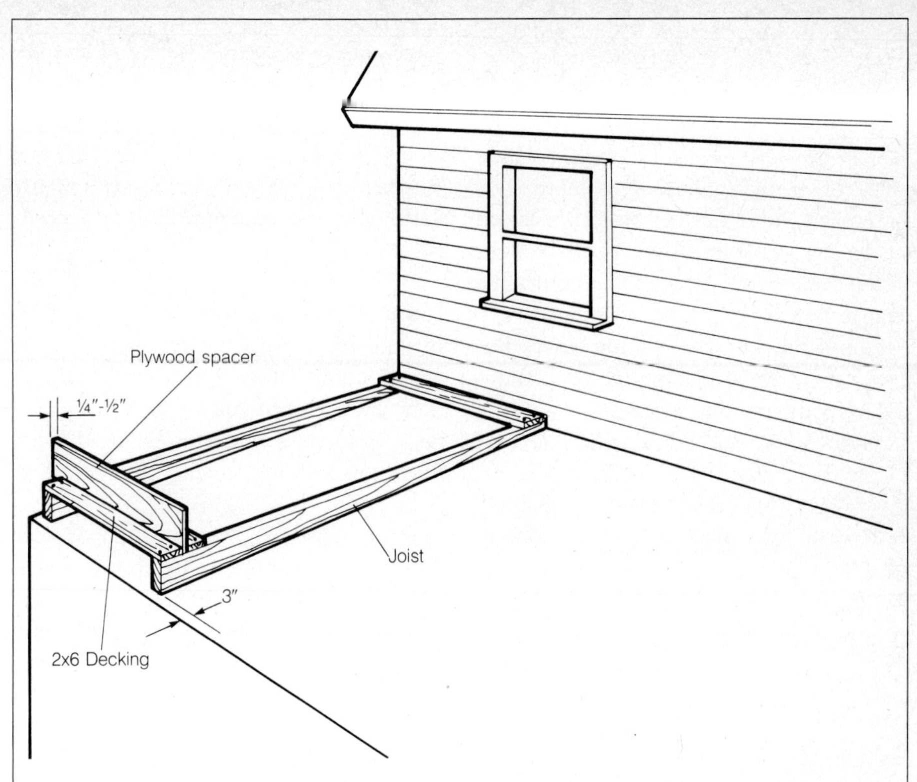

Build deck a unit at a time, attaching first two boards at either end of the foundation boards. Use a spacer to maintain even placement of boards along the run of the unit.

glass-panel door to provide light to the room.

Finding the studs. First locate the studs in the wall. These should be every 16 inches on center, but you may find that they are closer together or farther apart. You may be able to tell the approximate location by studying the nailing pattern of the baseboard. The base molding and trim usually nail into the studs. If there is no apparent pattern to the nailing, try to find the studs by knocking on the wall, listening for alternating hollow and solid sounds.

When you have found the studs, you will be able to tell if the window is framed against supporting studs or extra studs. If it seems to be framed between regularly spaced studs, then the framing space will probably be wide enough for a practical door.

Removing existing materials. Remove the window by prying off the interior facing and molding. Lift off the window sill. Remove any nails or screws that hold the frame in place.

Remove the plaster and lath or gypsum wallboard below the window. Cut it away until it is even with the wallboard that was under the sides of the window frame. You can now see the depth of the exterior siding. Use a circular saw set to the thickness of the siding to cut through that material. Remove the siding and any insulation below the window to the centerline of the framing studs on either side of the opening.

Resizing the opening. Measure the height of the existing opening. You will need a final opening high enough for the door you will be using and the thickness of the deck framing, deck surface and the threshold so the door will clear the deck surface easily. If the existing opening is not high enough, you must remove the existing doubled headers over the window opening, cut the cripple stud to size and install a new doubled header.

If you must remove this much of the wall supports, prop the ceiling first. Do this with two 2x4s shimmed top and bottom with ¼ inch plywood sheets to protect the ceiling and flooring from dents and scars. Leave the props in place until all reframing has been completed.

Information for rough openings comes with prehung door units. Purchase a door after you have determined where you want the door to go. Always allow for the height of the threshold when framing the opening.

Cutting where there is no window. If the attached garage is on a side of your house that has no windows, you will have to cut an entirely new opening through the wall. To do this, locate the studs, determine the desired location of the door and cut away the plaster or wallboard. Locate the door so it will be between studs so that you will have to cut out only one stud. Remove the wallboard or plaster from the center of the stud on one side of the door location to the center of the stud on the other. Measure up by the height of the door rough opening, plus the allowance for the threshold, and mark the center stud. Add to this the thickness of the doubled header that must be installed. Since exterior walls are load-carrying, prop the ceiling as described above before removing the stud.

Framing In the Door

If you have removed a window that has filled the space between studs, there should be a rough opening approximately 30 inches (32 inches from stud center to stud center.) There should also be a doubled header across the top of the opening with a cripple (partial) stud above the header. If the window opening is high enough to accommodate a door, you must: (1) remove the framing below the opening; (2) add new side trimmer studs; (3) install a new threshold so the door will be above the finished deck level; (4) slip

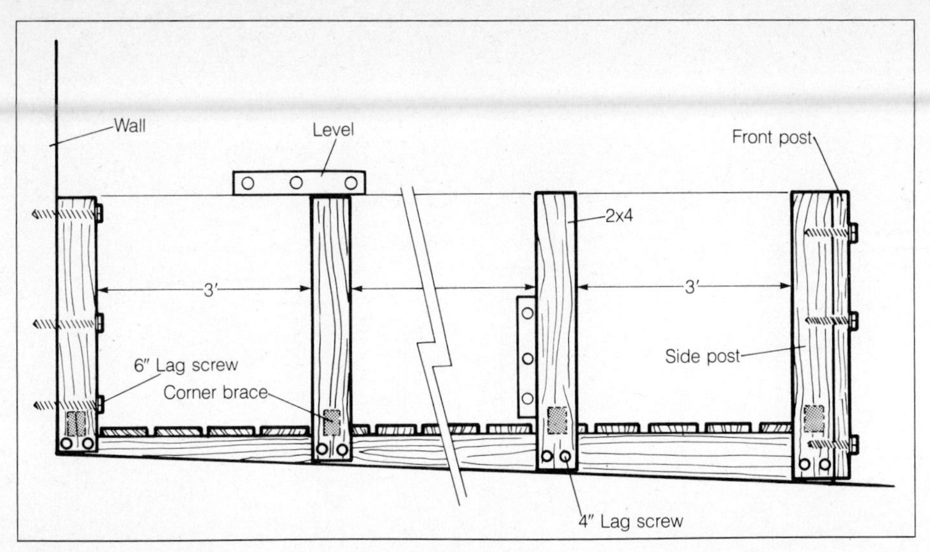

Posts for railings hold deck units together. Attach with lag screws into predrilled pilot holes. Be sure that you maintain level throughout the sections.

the door in place; (5) shim to level and plumb; (6) secure the door.

If your window was too narrow for a usable door, you will have to adjust the spacing of the studs. Add studs as needed for necessary wall support and to give the proper frame width.

If you had to cut through a solid wall, you need to cut through and remove the center stud. The length that you remove will be the rough opening height, the height of the finished threshold, plus the thickness of two 2x4s used as the doubled header. Measure carefully, checking the manufacturer's information and the actual thickness of the lumber. Toenail header boards to the side studs and into the cripple stud.

Installing the door

When the framing is complete, slip the new door in place. Check for level and plumb. Shim as needed before permanently mounting. Follow manufacturer's instructions. Finish the inside and outside with facings to match inside and outside trim. Add a drip cap flashing over the door trim. This may be either preformed flashing or a section of sheet metal that you bend yourself. The top fits under the siding; the bottom fits on and over the top of the door framing.

Building the Deck

Now that you can move in and out of the house easily, you can build the deck itself. Because you will be laying the deck over a roof, and because decks are usually—and reasonably—built with openings between the boards for quick drainage, construct so that sections of the deck may be removed easily to permit later maintenance work on the roofing material. The deck sections may be built as a series of modular units, held together by screws through the railing supports. Railing supports also attach to the house.

If the slope of the roof is only 1/4 inch per foot, your deck can follow that slope. However, if the pitch is greater than that slight slope, you will have to compensate for the slope by providing a deck base built of wedge-shape boards.

Planning the construction. Measure the area that you will be using for your deck. A single-car garage roof will provide approximately 10x20 feet of area. Allow an area 3 inches wide between the outside edges of your deck and the edge of the roof. This way, any rainfall dripping off the railing will fall on the roof and run into your gutter/downspout system rather than dripping directly to the ground.

It will be easier and more economical for you to build the deck in sections between 30 and 48 inches wide, but you may choose to build smaller sections if you are working alone. If you choose to build your deck the full size of a 10x20 roof area, you can construct six modular units 9 feet 9 inches long by 39 inches wide (10 feet minus 3 inches; and, 20 feet minus 6 inches divided by 6). You may also build 13 units each 18 inches wide, if you think this will be easier to handle. You may decide to deck only part of the roof to create a 10x12 deck. Note that any modular unit that is over 28 inches wide will require a center support board. Units

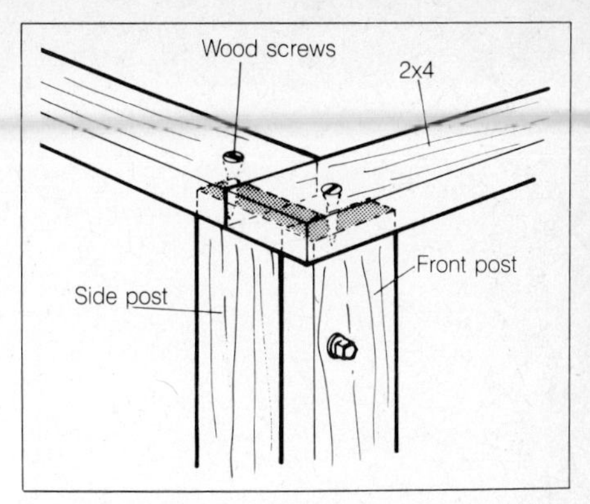

Center 2x4 railing on posts and secure with wood screws. Additional railings can be placed between deck level and top railings.

28 inches or less in width may be built with only side rail supports.

Measuring the slope. Lay a 2x4 (or thicker board) on the roof, perpendicular to the house wall. The board must be at least as long as the roof is wide. Lay the board widest side down. Place a level on the board and begin lifting the end near the edge until the bubble in the level is centered. Measure the distance from the roof to the level board. If the slope of your roof is 1/2 inch per foot, this will be approximately 5 inches.

If you are using 2x4s or 2x6s for your decking and you lay the boards flat, the decking thickness will be 1 1/2 inches. If you have set your access door so that the bottom clears the roof surface by 4 inches, you will have to cut the decking support (foundation) boards so that the end nearest the door will be 2 inches thick. The other end will be 7 inches thick to bring the deck to an even level. You will be able to cut this from 2x8 stock 10 feet long.

Marking and cutting the boards. Lay your boards flat and measure off the width of the roof less three inches. Assuming a 10 foot width, this will be 9 feet 9 inches. Cut the boards at this mark. At one end, measure down 2 inches from the top edge; at the other end, measure down seven inches. Draw a straight line across the face of the board between these two marks. Cut along the line using a circular saw or reciprocating saw. You will need 18 boards—three for each of the six modular units.

Constructing the modules. To build a 39-inch-wide unit, lay out three of your foundation boards so that they are evenly spaced, 39 inches from outside edge to outside edge. The middle board will be located exactly on a 19 1/2 inch center from

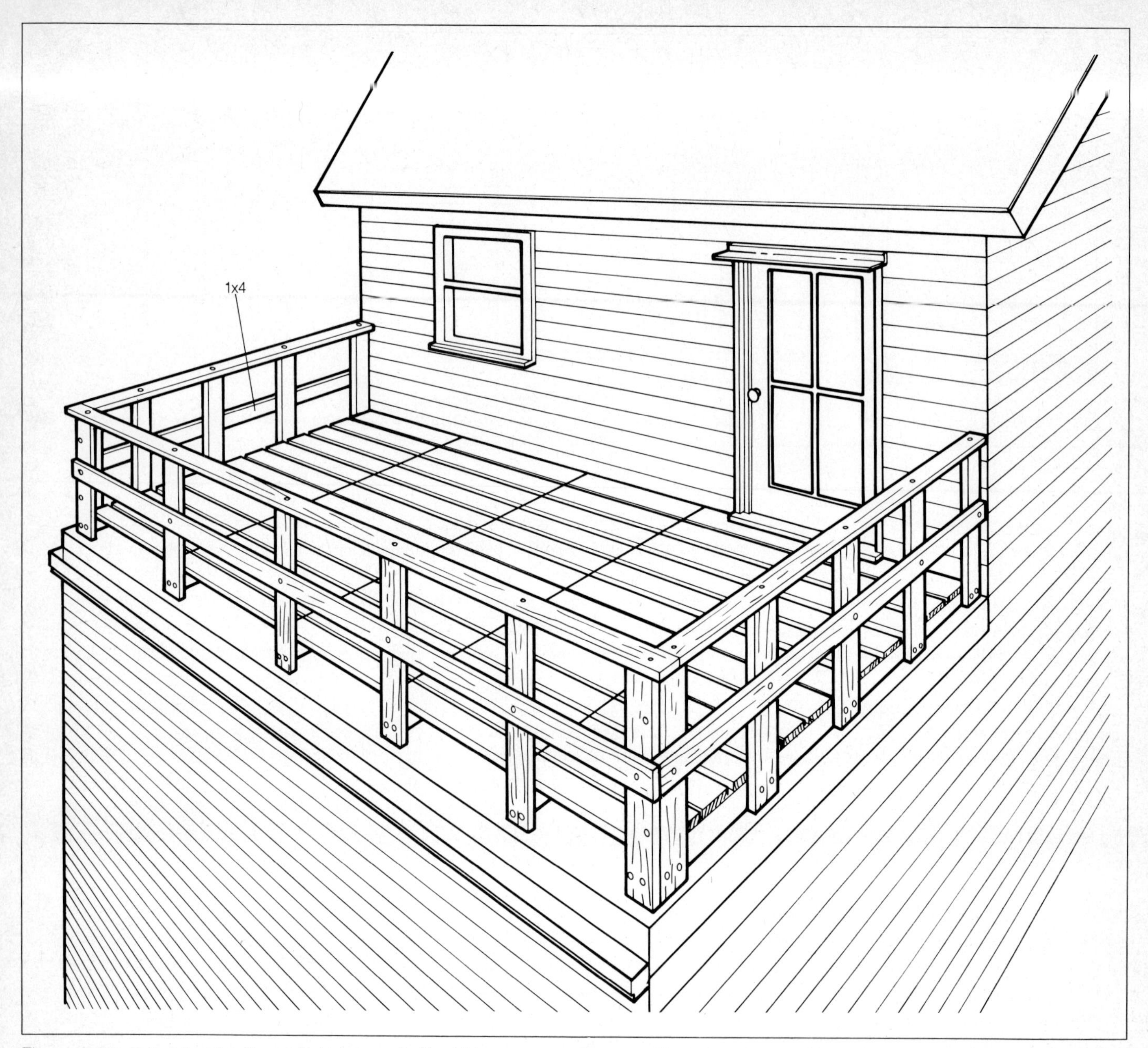

The completed deck is attractive, safe to use, and will add to your recreation area. It can be taken apart for roof maintenance.

the outside edges of the side boards. Using 10d galvanized nails, nail one 2x4 or 2x6 piece of decking flush with each end to hold the foundation boards in position. The decking boards should be cut to fit the exact width of the modules. Allow approximately 1/4 inch space between boards for drainage. It will take 31 2x4s or 20 2x6s to deck each module with approximately 1/4 inch spacing. This will give you 3/4 inch of excess space to distribute along the run if you are using 2x4s. You will have 2 inches of excess spacing to distribute over the run if you are using 2x6 boards for the decking.

Lay out the decking boards using a 1/4 inch spacer, a piece of 1/4 inch plywood will do, then adjust the spacing to absorb the excess and mark the positions of the decking boards on the foundation boards. Nail the decking in place with 10d galvanized nails. Repeat for each modular unit needed. Set units in place.

Adding the railings. Cut 2x4 boards to a length of between 43 and 45 inches, depending on the height you want for your top railing. For a 10x20 foot deck, you will have to cut 15 railing pieces. Two boards are attached to the house at either side of the deck. Use 6-inch lag screws driven through the narrow face of the board then into the deck foundation boards. Evenly space railing supports no more than three feet on center. Level and plumb along each side. Attach one board flush with the edge at the corner. Butt this rail support board with the board on the other side of the corner. Attach these boards with lag screws to each other and to the foundation boards. Install one railing support at each joint between modules so that the modules are held together by lag screws through the foundation boards. Check each railing support board carefully for level and plumb. Always keep the spacing under three feet on center; add more boards as needed. Repeat corner and side installation for the other side.

Center 2x4 railings on the posts and attach with galvanized wood screws. Connect posts with additional bracing. Check codes for minimum construction standards.

A GLOSSARY OF ROOFING TERMS

Alignment notch Cutout projection or slit on the ends or sides of shingles. Acts as a guide in application to secure proper exposure.

Asphalt A bituminous compound, dark brown or black in color, used in the manufacture of asphalt roofing shingles.

Attic The space immediately under the sloping roof of a house.

Barrel vault Simplest form of arched roof, consisting of a continuous arch of semi-circular sections.

Base flashing That portion of the flashing which is attached to, or rests on, the roof deck to direct the flow of water on the roof away from the juncture of the deck and the parapet wall.

Blends Mixtures of various colored granules found on the one face of mineral-surfaced roofing.

Blind nailing Nails driven so that the heads are concealed.

Bond A roofing bond is a guarantee often furnished to the owner, generally by a bonding company, for the manufacturer of the roofing materials to maintain a specified roof in weathertight condition for a specified time. Terms and conditions of individual bonds vary from manufacturer to manufacturer.

Boston lap A method of finishing the ridge of a shingle course, using overlapping vertical joints.

Bowstring roof Constructed with curved timber trusses and horizontal tie-beams connected by light diagonal lattices of wood.

Built-up roofing An outer covering of a comparatively flat roof, consisting of several layers of saturated and/or saturated-and-coated felt, each layer mopped with hot tar or asphalt as laid, and the top layer finished with mineral or rock covering, or with a special coating.

Butt That portion of a shingle exposed to the weather, generally called the tab of the shingle.

Cantilever A self-supporting projection without external bracing in which a beam or series of beams is supported by a downward force behind a fulcrum.

Cant strip A beveled wood, fiberboard, metal, gypsum or concrete strip at the juncture of the roof with vertical surfaces to break the acute angle.

Cap flashing The portion of the flashing that is built into a vertical surface to prevent water seepage behind the base flashing. Cap flashing overlaps the base flashing.

Caulk To fill or seal a joint with mastic or cement.

Cement A substance which, by curing between the two surfaces to which it has been adhered, binds them together.

Ceramic granules Roofing granules in which color is fused to rock under extreme heat to provide a long lasting finish.

Chicken ladder Hooks over the ridge by means of broad 2x4s nailed to the top, to provide safe footing on steep pitches.

Chimney Masonry or brick work containing one or more flues, projecting through and above the roof.

Clerestory An upward extension of enclosed daylighted space created by carrying a setback vertical, windowed wall up and through the roof slope.

Clipped gable A gable cut back at the peak in a hip-roof form.

Closed valley A valley in which the roofing material is laced or woven through the valley intersection.

Coating Hard, heavy, viscous, asphalt layer between the saturated fiber base and granule surface on roofing materials.

Code or building code Legal restrictions of a given locality, governing construction of buildings and methods and materials used in construction.

Collars or vent sleeves Sheet metal flanged collars placed around vent pipes for the purpose of sealing off the roofing around the vent pipe openings.

Conical-style flashing A cone-shaped section that fits over prefabricated chimney sections that pass through a steeply pitched roof. The conical flashing fits under upper-level shingles and over lower-level shingles. The cone is adjustable to fit various degrees of pitch. The cone also directs rain away from the chimney and seals the roof opening.

Cornice A projecting horizontal feature located at or near the top of a wall, used when there is no overhang. A horizontal, decorative projection at the top of a building.

Counter flashing Strips of metal, roofing, or fabric inserted and securely anchored in the reglet or attached to a vertical surface above the plane of the roof and turned down over the base flashing to protect the base flashing.

Couped roof Constructed without ties or collars, its rafters being fixed to the wall plates and ridge pieces.

Course A horizontal unit of roofing shingles running the length of the roof.

Coverage Term applied to indicate roughly the amount of weather protection provided by different roofing products, relating to the amount of materials used. Coverage is determined on the basis of coverage patterns and design.

Coverage patterns Pattern of application for roofing materials, as related to the amount of cover or overlap.

Cricket A small, peaked saddle constructed behind the chimney to prevent accumulation of snow and ice and to deflect draining water around the chimney.

Cupola A structure, square to round in plan, rising above a main roof. While generally ornamental, a cupola also enables ventilation of an attic.

Cutout The portion of a strip shingle cut out to produce the tab to give the effect of individual shingles. Sometimes referred to as a slot or a notch.

Dead level A roof without slope.

Dead load The total weight of all installed materials, and the constant weight of a roof, used to compute the strength of all supporting framing members.

Deck The material first installed over framing members and over which roofing material is applied. It usually is solid plywood sheathing, but may be of spaced boards.

Double coverage Method of applying roof shingles so that two complete layers of protection are provided. New construction requires double coverage shingles.

Dormer A window unit projecting through the sloping plane of a roof.

Down spout A pipe used for draining water from the roof.

Drip edge A noncorrosive, nonstaining metal or vinyl strip used along the eaves and rakes, designed to allow water run off to drip free of underlying construction.

Drip course The first course of shingles at the eave.

Eaves The edge of a roof that projects over the outside wall.

Eave trough A gutter along the eave of the roof.

Ell An extension of a building, set at right angles to its length.

Exposure The portion of a shingle that is exposed to the weather. Exposure is usually measured from the butt of one shingle to the butt of the next overlying shingle.

Eyebrow A dormer, usually of small size, whose roof line over the upright face is an arch curve, turning into a reverse curve to meet the horizontal at either end.

Facade The face of a building, usually the front.

Fascia A horizontal band or vertical face, usually below the edge of the roof to finish off rafter ends.

Field The area covered by a roofing material.

Fire-rated asphalt roofing Rock/ceramic surfaced asphalt shingle roofing that has been tested and rated for fire resistance by Underwriters' Laboratories, Inc.

Fire resistant Descriptive of building materials that resist destruction by fire for a definite duration of time.

Fire brands Flaming pieces of material or burning embers.

Flashing Material used to prevent seepage of wind or water around any intersection or projection in a roof, including vent pipes, chimneys, adjoining walls, dormers, and valleys.

Gable The upper part of a wall under the ridge of a pitched roof; the end or wing of a building.

Gambrel A double pitched roof that terminates in a small gable at the ridge where the angle of pitch is abruptly changed between ridge and eave.

Granules Finely ground or crushed rock that is ceramic-coated and used on the exposed surface of roofing products. Granules are available in various sizes and colors.

Guarantee Usually refers to the assurance given by the roofing subcontractor that the work performed by him is in accordance with specification and is without defects in labor and material.

Gutter A channel for water at a roof edge or at ground level.

Head lap The shortest distance from the exposed butt or tab of the overlying shingle, or from the top of the cutout in the overlying shingle to the top edge of the underlying shingle.

Helm roof A steeply pitched roof that has four faces converging at the top, with a gable at the foot of each.

Hexagonal or hex A strip shingle having a butt or tab, which is one-half of a hexagon.

Hip The line of intersection of two roof planes, the eave lines of which are not parallel.

Hip rafter A rafter used to form the hip of a roof.

Hip roof A roof with sloped planes instead of vertical ends.

Hyperbolic paraboloid roof A special form of double-curved shell, the geometry of which is generated by straight lines. The shape consists of a continuous plane developing from a parabolic arc in one direction to a similar inverted parabola in the other direction.

Incline *See Pitch, Slope.*

Jack rafter Any one of the shorter rafters used from the plate to the angle rafter of a hip roof.

Lacing or Weaving Interweaving of a course of shingles where there is an intersection in a roof; e.g., at 90° angles in a valley.

Laid to the weather *See Exposure.*

Lap To overlap the surface of one shingle with another; and length of such overlap.

Lean-to roof Has one slope only and is built against a higher wall.

Live load The total weight of all installed equipment and materials and all variable weight (such as snow, ice and people) that will move across a surface. Used to compute the strength of all supporting framing members.

Lock shingles Designed with a mechanical locking feature to provide effective wind resistance.

Mansard roof A roof having a double slope; the lower one is longer and steeper. Named for French architect Francois Mansard (1598–1666).

Monitor A continuous section of roof raised to admit light on a vertical plane.

Open valley When the roofing material is not laced or woven at the valley intersection, but trimmed so the flashing material is exposed.

Parapet A low retaining wall found at the edge of a roof, porch or terrace.

Reglet A groove in the vertical wall adjacent to a roof surface above the top of base flashing, and into which the metal counter flashing is placed and rigidly held in place; it is either formed in concrete or consists of a metal insert, or a "reglet block" of masonry.

Ridge The top horizontal member of a sloping roof, against which the upper ends of the rafters are fixed.

Roll Roofing Roofing material laid and overlapped from a roll of material.

Roof The weatherproof top shell of a building.

Roofers cement A quick-setting asphalt adhesive for use with roofing materials.

Roof-tree The ridge of a roof.

Roof truss Any type of truss used for roof support.

Run The horizontal distance from the eave to a point directly under the ridge of a roof.

Saddle The ridge in a small roof deck that divides the roof in order to divert water to the drain.

Peaked roof A roof rising either to a point or a ridge.

Penthouse Enclosed space above the level of a main flat roof, as at the top of an elevator shaft or above-roof apartment.

Pitch Height from the joist to the ridge, divided by the rafter length; this equates to rise in inches per horizontal foot.

Plate *See Wall plate.*

Pointing trowel A trowel used in tuckpointing mortar. Tuckpointing is a repair method in which mortar is dug out of joints and replaced.

Portico A roofed space, open or partly enclosed, forming the entrance and centerpiece of the facade of a house.

Principals The main rafters of a roof, usually corresponding to the main bay divisions of the space below.

Purlin A horizontal member resting usually on trusses and supporting the roof rafters.

Rafter A supporting member immediately beneath the roof deck sloping from the wall plate to the ridge.

Rake A slope or inclination of a roof; the same as the slope.

Saddleback roof A normal pitched roof, usually for roofs of towers.

Safety harness Either a manufactured leather strap vest and rope unit, or a looped and tied rope vest and safety line.

Saturant Asphalt used to impregnate felt to waterproof it and give it strength.

Seal down or Positive sealdown A powerful asphalt adhesive factory-applied so that the shingles, once installed, have a concealed strip of sealing compound that securely bonds each shingle to the one above to provide excellent wind resistance.

Self spacing Notch or small tab at one or both ends, at sides of shingles; when engaged, interlocked or butted together it provides uniform spacing.

Shingle Material applied in overlapping courses to cover a roof.

Side lap The horizontal distance that the shingle overlaps the adjacent shingle in the same course; also the horizontal distance one sheet of roofing overlaps an adjacent sheet.

Single coverage Method of applying roof shingles to provide only one complete layer of roof protection. Many special shingles for reroofing are designed for single coverage for reasons of economy, weight, and flexibility.

Skylight A glazed opening in the roof permitting light and/or ventilation.

Slope The degree of inclination of a roof plane in inches of rise per horizontal foot.

Soffit The finished underside of an eave.

Soil stacks *See Vents.*

Span The horizontal measurement from eave to eave.

Spire A tall tower roof that tapers upward to a point.

Square An area of exposed roofing 10 feet square, or comprising 100 square feet.

Square butt shingles Strip shingles that usually have two or three tabs formed by cutouts or slots.

Starter course The first course of shingles installed on a roof, starting at the lower edge of the eave. It is covered by the first course.

Starter strip Mineral-surfaced roll roofing applied at eave line before application of shingles to fill spaces of cutouts and joints. May also be made by cutting off tabs of shingles on job-site, which is then called a starter course.

Steeple A tower and spire, usually on a church.

Stepped flashing Flashing along a roof slope against a wall or chimney, where the masonry joint entered must necessarily change at intervals to keep a safe distance above the roof surface. The flashing goes under the roof siding and the shingles. It usually consists of L-shaped units that fit into the joint between the roof plane and the vertical obstacle.

Storm collar A flashing unit for prefabricated chimney pipe. The collar is designed to direct heavy precipitation away from the chimney roof joint.

Tab Portion of strip shingles defined by cutouts or slots, so that when installed, material appears to be individually applied.

Tabbing Method of applying high strength adhesives to shingles for wind resistance.

Tarbuck knot A safety knot designed for use by mountain climbers. The Tarbuck knot is used in creating a rope safety harness.

"UL" label Underwriters' Laboratories, Inc. is the most widely known and accepted testing agency for testing fire-resistance properties of materials and constructions.

Underlayment An asphalt-saturated or coated felt applied over the roof deck and under the roofing material.

Valley The line of intersection of two roof slopes.

Vent An outlet for air; vent pipe in a plumbing system; a ventilating duct.

Vent sleeves or Collars Sheet-metal-flanged collars placed around vent pipes to seal off the roofing around the vent pipe opening.

Wall plate A timber laid longitudinally on the top of a wall to cover the ends of the rafters.

Weather To undergo the changes in color, texture or efficiency brought about by continued exposure to wind, rain, sun, frost, snow, and other elements.

Weaving or Lacing Interweaving of a course of shingles where there is an intersection in a roof for drainage (e.g., valley).

Zoning Restriction as to size or character of buildings permitted within specific areas, as established by urban authorities.

Glossary courtesy Celotex Corporation

PRODUCT SOURCES

ASPHALT SHINGLES

Asphalt Roofing Manufacturers Assn.
c/o Summer, Rider & Associates
355 Lexington Avenue
New York, NY 10017

Bird & Son inc.
Washington St.
E. Walpole, MA 02032

Bird & Son, inc.
6600 S. Central Ave.
Chicago, IL 60638

Celotex Corp.
1500 N. Dale Mabry
Tampa, FL 33607

Certain Teed Corp.
P.O. Box 860
Valley Forge, PA 19482

Flintkote Building Materials
20200 Governors Drive
Olympia Fields, IL 60461

GAF Building Products Div.
140 W. 51st St.
New York, NY 10020

Globe Industries Inc.
2638 E. 126th St.
Chicago, IL 60633

Johns-Manville Co.
Greenwood Plaza
Denver, CO 80217

ATTIC VENTILATORS

Arvin Industries, Inc.
Columbus, IN 47201

The G.C. Breidert Co.
P.O. Box 1190
San Fernando, CA 91341

Exitaire Co.
P.O. Box 276
Pacoima, CA 91330

Kool-O-Matic Corp.
1831 Terminal Road
Niles, MI

Leigh Products Inc.
411-64th Avenue
Coopersville, MI 49404

Triangle Engineering Company
P.O. Drawer 38271
Houston, TX 77088

CAULKS, CEMENTS, SEALANTS

Abitibi Building Products Div.
3250 W. Big Beaver Rd.
Troy, MI 48084

W.R. Bonsal Company
P.O. Box 241148
Charlotte, NC 28224

D-A-P Inc.
P.O. Box 277
Dayton, OH 45401

Dow-Corning Corp.
2200 W. Salzburg Rd.
Midland, MI 48640

3 E Corp.
P.O. Box 177
Somerdale, NJ 08083

Hadley Adhesives, Div. Sherwin Williams
2827 Breckenridge Industrial Court
St. Louis, MO 63144

W.R. Meadows Corp.
12 Kimball St.
Elgin, IL 60120

Miracle Adhesives Corp.
250 Pettit Ave.
Bellmore, NY 11710

Pecora Corp.
165 Wambold Rd.
Harleysville, PA 19438

Thoro System Products
7800 NW 38th St.
Miami, FL 33166

Tremco Mfg. Co.
10701 Shaker Blvd.
Cleveland, OH 44104

CEDAR SHINGLES AND SHAKES

J.H. Baxter Co.
1700 South El Camino Real
San Mateo, CA 94402

Evans Products Co.
P.O. Box 3295
Portland, OR 97208

Koppers Forest Products Div.
750 Koppers Bldg.
Pittsburgh, PA 15219

Red Cedar Shingle & Handsplit Shake Bureau
5510 White Building
Seattle, WA 98101

Shakertown Corp.
P.O. Box 400
Winlock, WA 98596

FANCY-BUTT SHINGLES

Homestead Mills Ltd.
3247 63rd St., SW
Seattle, WA 98116

Koppers Co.
Koppers Building
Pittsburgh, PA 15219

Red Cedar Shingle & Handsplit Shake Bureau
5510 White Building
Seattle, WA 98101

Shakertown Corp.
P.O. Box 400
Winlock, WA 98596

FANCY-BUTT SHINGLES

Koppers Co.
Koppers Building
Pittsburgh, PA 15219

Red Cedar Shingle & Handsplit Shake Bureau
5510 White Building
Seattle, WA 98101

Shakertown Corp.
P.O. Box 400
Winlock, WA 98596

FASCIA-SOFFIT SYSTEMS

Alcan Building Products
Division of Alcan Aluminum Corporation
280 North Park Avenue
Warren, OH 44481

Alcoa Building Products Inc.
Two Allegheny Center
Pittsburgh, PA 15212

Aluminum Trim & Shapes Corp.
30 Deep Rock Rd.
Rochester, NY 14624

Bird & Son inc.
E. Walpole, MA 02032

CertainTeed Corp.
P.O. Box 860
Valley Forge, PA 19482

Crown Aluminum Div. Whittaker
P.O. Box 61
Roxboro, NC 27573

Edco Products Inc.
825 Excelsior Ave. East
Hopkins, MN 55343

GAF Building Products Div.
140 W. 51st St.
New York, NY 10020

General Aluminum Corp.
Box 200 Baseline Rd.
Montgomery, IL 60538

Norandex Building Materials
7120 Krick Rd.
Cleveland, OH 44146

Omni Products Co.
1540 W. Fullerton Ave.
Addison, IL 60101

Rollex Corp.
2001 Lunt Ave.
Elk Grove, Village IL 60007

Superior-Vydel Co.
1660 Old Deerfield Rd.
Highland Park, IL 60035

Upson Co.
Upson Point
Lockport, NY 14094

FASTENING TOOLS

Arrow Fastener Co.
271 Mayhill St.
Saddle Brook, NJ 07662

Bostitch Div, Textron
117 Briggs Drive
East Greenwich, RI 02818

Duo-Fast Fastener Corp.
3702 N. River Rd.
Franklin Park, IL 60131

Paslode Div. Signode
8080 McCormick Blvd.
Skokie, IL 60076

Senco Products Inc.
8485 Broadwell Rd.
Cincinnati, OH 45244

Swingline Consumer Products
32-00 Skillman Ave.
Long Island City, NY 11101

HAND TOOLS

Diamond Tool Co.
P.O. Box 6246
Duluth, MN 55806

Goldblatt Tool Co.
P.O. Box 2334
Kansas City, KS 66110

Oxwall Tool Co., Inc.
133-10 32nd Ave.
Flushing, NY 11352

Stanley Tools Div.
no-street-needed
New Britain, CT 06050

INSULATION MATERIALS

CertainTeed Corp.
P.O. Box 860
Valley Forge, PA 19482

W.R. Grace Construction Products Div.
62 Whittemore Ave.
Cambridge, MA 02140

Johns-Manville Co.
Greenwood Plaza
Denver, CO 80217

Owens-Corning Fiberglas Corp.
Fiberglas Tower
Toledo, OH 43659

Rockwool Industries Inc.
P.O. Box 5170
Denver, CO 80217

LADDERS, LADDER DEVICES

Alproco Inc.
P.O. Box 863
Melbourne, FL 32935

Fracon Co.
690 Wellesley St.
Weston, MA 02193

Goldblatt Tool Co.
P.O. Box 2334
Kansas City, KS 66110

Louisville Ladder Div.
1163 Algonquin Ave.
Louisville, KY 40208

R.D. Werner Co.
P.O. Box 580
Greenville, PA 16125

POWERED VENTILATORS

Arvin Industries, Inc.
Columbus, IN 47201

Butler Vent-A-Matic Corp.
P.O. Box 728
Mineral Wells,TX 76067

Cool Attic Corp.
P.O. Box 11558
Ft. Worth, TX 76109

Emerson Environment Products
Division of Emerson Electric Co.
8400 Pershall Road
Hazelwood, MO 63042

Home Ventilating Institute
4300-L Lincoln Avenue
Rolling Meadows, IL 60008

Kool-Matic Corp.
1831 Terminal Rd.
Niles, MI 49120

Leigh Products Inc.
411-64th Avenue
Coopersville, MI 49404

NuTone Div. Scoville
Madison & Red Bank Rds.
Cincinnati, OH 45227

Phil Rich Fan Mfg.
P.O. Box 55589
Houston, TX 77055

RAIN-CARRYING SYSTEMS

Alcan Aluminum Building Products
Division of Alcan Aluminum Corporation
280 North Park Avenue
Warren, OH 44481

Alcoa Building Products Inc.
Two Allegheny Center
Pittsburgh, PA 15212

Bird & Son, inc.
East Walpole, MA 02032

Crown Aluminum Div. Whittaker
P.O. Box 61
Roxboro, NC 27573

Genova, Inc.
Raingo Systems
7034 East Court St.
Davison, MI 48423

Inryco, Inc.
Milcor Div.
P.O. Box 393
Milwaukee, WI 53201

Kaiser Aluminum & Chemical Sales
5201 Enterprise Blvd.
Toledo, OH 43612

Nichols Homeshield Inc.
1000 Harvester Rd.
West Chicago, IL 60185

Norandex Aluminum Building Products
7120 Krick Rd.
Cleveland, OH 44146

Revere Aluminum Building Products
11440 W. Addison St.
Franklin Park, IL 60131

Rollex Co.
2001 Lunt Ave.
Elk Grove Village, IL 60007

SCAFFOLD DEVICES

Dalton Manufacturing Co.
130 S. Bemiston Ave.
St. Louis, MO 63105

Hoitsma Adjustable Scaffold Bracket Co.
P.O. Box 595
Paterson, NJ 07544

Metal Fabricators Inc.
P.O. Box 752
Ellwood City, PA 16117

SKYLIGHT MATERIALS

APC Corporation
44 Utter Avenue
Hawthorne, NJ 07506

Keller Companies Inc.
P.O. Box 327
Manchester, NH 03105

Skymaster Div.
413 Virginia Drive
Orlando, FL 32803

Velux-America Inc.
74 Cummings Park
Woburn, MA 01801

Ventarama Skylight Corporation
75 Channel Drive
Port Washington, NY 11050

Wasco Products Inc.
P.O. Box 351
Sanford, ME 04073

Metric Conversions

LUMBER

Sizes: Metric cross-sections are so close to their nearest Imperial sizes, as noted below, that for most purposes they may be considered equivalents.

Lengths: Metric lengths are based on a 300mm module which is slightly shorter in length than an Imperial foot. It will therefore be important to check your requirements accurately to the nearest inch and consult the table below to find the metric length required.

Areas: The metric area is a square metre. Use the following conversion factors when converting from Imperial data: 100 sq. feet = 9.290 sq. metres.

METRIC SIZES SHOWN BESIDE NEAREST IMPERIAL EQUIVALENT

mm	Inches	mm	Inches
16 x 75	5/8 x 3	44 x 150	1¾ x 6
16 x 100	5/8 x 4	44 x 175	1¾ x 7
16 x 125	5/8 x 5	44 x 200	1¾ x 8
16 x 150	5/8 x 6	44 x 225	1¾ x 9
19 x 75	¾ x 3	44 x 250	1¾ x 10
19 x 100	¾ x 4	44 x 300	1¾ x 12
19 x 125	¾ x 5	50 x 75	2 x 3
19 x 150	¾ x 6	50 x 100	2 x 4
22 x 75	7/8 x 3	50 x 125	2 x 5
22 x 100	7/8 x 4	50 x 150	2 x 6
22 x 125	7/8 x 5	50 x 175	2 x 7
22 x 150	7/8 x 6	50 x 200	2 x 8
25 x 75	1 x 3	50 x 225	2 x 9
25 x 100	1 x 4	50 x 250	2 x 10
25 x 125	1 x 5	50 x 300	2 x 12
25 x 150	1 x 6	63 x 100	2½ x 4
25 x 175	1 x 7	63 x 125	2½ x 5
25 x 200	1 x 8	63 x 150	2½ x 6
25 x 225	1 x 9	63 x 175	2½ x 7
25 x 250	1 x 10	63 x 200	2½ x 8
25 x 300	1 x 12	63 x 225	2½ x 9
32 x 75	1¼ x 3	75 x 100	3 x 4
32 x 100	1¼ x 4	75 x 125	3 x 5
32 x 125	1¼ x 5	75 x 150	3 x 6
32 x 150	1¼ x 6	75 x 175	3 x 7
32 x 175	1¼ x 7	75 x 200	3 x 8
32 x 200	1¼ x 8	75 x 225	3 x 9
32 x 225	1¼ x 9	75 x 250	3 x 10
32 x 250	1¼ x 10	75 x 300	3 x 12
32 x 300	1¼ x 12	100 x 100	4 x 4
38 x 75	1½ x 3	100 x 150	4 x 6
38 x 100	1½ x 4	100 x 200	4 x 8
38 x 125	1½ x 5	100 x 250	4 x 10
38 x 150	1½ x 6	100 x 300	4 x 12
38 x 175	1½ x 7	150 x 150	6 x 6
38 x 200	1½ x 8	150 x 200	6 x 8
38 x 225	1½ x 9	150 x 300	6 x 12
44 x 75	1¾ x 3	200 x 200	8 x 8
44 x 100	1¾ x 4	250 x 250	10 x 10
44 x 125	1¾ x 5	300 x 300	12 x 12

METRIC LENGTHS

Lengths Metres	Equiv. Ft. & Inches
1.8m	5′ 10⅞″
2.1m	6′ 10⅝″
2.4m	7′ 10½″
2.7m	8′ 10¼″
3.0m	9′ 10⅛″
3.3m	10′ 9⅞″
3.6m	11′ 9¾″
3.9m	12′ 9½″
4.2m	13′ 9⅜″
4.5m	14′ 9⅓″
4.8m	15′ 9″
5.1m	16′ 8¾″
5.4m	17′ 8⅝″
5.7m	18′ 8⅜″
6.0m	19′ 8¼″
6.3m	20′ 8″
6.6m	21′ 7⅞″
6.9m	22′ 7⅝″
7.2m	23′ 7½″
7.5m	24′ 7¼″
7.8m	25′ 7⅛″

All the dimensions are based on 1 inch = 25 mm.

NOMINAL SIZE (This is what you order.)	ACTUAL SIZE (This is what you get.)
Inches	Inches
1 x 1	¾ x ¾
1 x 2	¾ x 1½
1 x 3	¾ x 2½
1 x 4	¾ x 3½
1 x 6	¾ x 5½
1 x 8	¾ x 7¼
1 x 10	¾ x 9¼
1 x 12	¾ x 11¼
2 x 2	1¾ x 1¾
2 x 3	1½ x 2½
2 x 4	1½ x 3½
2 x 6	1½ x 5½
2 x 8	1½ x 7¼
2 x 10	1½ x 9¼
2 x 12	1½ x 11¼

NAILS

NUMBER PER POUND OR KILO

Size	Weight Unit	Common	Casing	Box	Finishing
2d	Pound	876	1010	1010	1351
	Kilo	1927	2222	2222	2972
3d	Pound	586	635	635	807
	Kilo	1289	1397	1397	1775
4d	Pound	316	473	473	548
	Kilo	695	1041	1041	1206
5d	Pound	271	406	406	500
	Kilo	596	893	893	1100
6d	Pound	181	236	236	309
	Kilo	398	591	519	680
7d	Pound	161	210	210	238
	Kilo	354	462	462	524
8d	Pound	106	145	145	189
	Kilo	233	319	319	416
9d	Pound	96	132	132	172
	Kilo	211	290	290	398
10d	Pound	69	94	94	121
	Kilo	152	207	207	266
12d	Pound	64	88	88	113
	Kilo	141	194	194	249
16d	Pound	49	71	71	90
	Kilo	108	156	156	198
20d	Pound	31	52	52	62
	Kilo	68	114	114	136
30d	Pound	24	46	46	
	Kilo	53	101	101	
40d	Pound	18	35	35	
	Kilo	37	77	77	
50d	Pound	14			
	Kilo	31			
60d	Pound	11			
	Kilo	24			

LENGTH AND DIAMETER IN INCHES AND CENTIMETERS

Size	Length		Diameter	
	Inches	Centimeters	Inches	Centimeters*
2d	1	2.5	.068	.17
3d	1/2	3.2	.102	.26
4d	1/4	3.8	.102	.26
5d	1/6	4.4	.102	.26
6d	2	5.1	.115	.29
7d	2/2	5.7	.115	.29
8d	2/4	6.4	.131	.33
9d	2/6	7.0	.131	.33
10d	3	7.6	.148	.38
12d	3/2	8.3	.148	.38
16d	3/4	8.9	.148	.38
20d	4	10.2	.203	.51
30d	4/4	11.4	.220	.58
40d	5	12.7	.238	.60
50d	5/4	14.0	.257	.66
60d	6	15.2	.277	.70

*Exact conversion

INDEX

Acknowledgments, Contributors, Addresses, Picture Credits

We wish to extend our thanks to the individuals, associations, and manufacturers who graciously provided information and photographs for this book. Specific credit for individual photos is given below, with the names and addresses of the contributors. All photos not otherwise identified were provided by the author or the staff of Creative Homeowner Press.

Asphalt Roofing Manufacturers Association c/o Sumner Rider & Associates, 355 Lexington Avenue, New York, New York 10017. Information on 82–83 on roll roofing adapted with permission from ARMA. 36, 47 lower right, 52, 58 lower right, 59, 60, 61, 63 right upper four, 71, 72, 73, 76 upper right, 88, 91 upper; drawings on pages 66, 67, 70, 71, 72, 73, 74, 76, 81, 82, 83, 84, 85, 86, 87, 88, 89, 90, 91 adapted from ARMA by permission.

Bird & Son inc East Walpole, Massachusetts 02032. 6 upper left, 13, 16, 22 upper 6 photos

Michael Bliss, landscape architect, 221 Sunset Drive, Encinitas, California 92024. 49 upper right, 50, 63 upper left

James W. Brett, architectural photography, 1070 West Orange Grove, Tucson, Arizona 85704. 51

Monte Burch Humansville, Missouri 65674. 58 right center

Celotex Corp. 1500 North Dale Mabry, Tampa, Florida 33607. 76 lower left

Ego Productions, James M. Auer, Charles W. Auer, 1849 North 72nd Street, Wauwatosa, Wisconsin 53213. JMA: 55 lower, 56, 58 upper, lower left, 68, 120, 121 upper left and upper right. CWA: 7 upper left, 18, 20, 22 lower right, 23 lower right, 24

Georgia-Pacific Corporation 900 SW 5th Street, Portland, Oregon 97204. 69

Kool-O-Matic 1831 Terminal Road, Niles, Michigan 49120. 98 upper right

Koppers Co., Inc. Koppers Building, Pittsburgh, Pennsylvania 15219. 53, 54, 55 upper, 63 lower left

Lied's Green Valley Gardens N63 W22039 Highway 74, Sussex, Wisconsin 53089. 63

Leigh Products Inc. 411-64th Avenue, Coopersville, Michigan 49404. 98 center and lower right

Richard V. Nunn Media Mark Productions, Falls Church Inn, 6633 Arlington Blvd., Falls Church, Virginia 22045. 7 upper center, 21, 39, 41, 44, 47 upper center, 48, 80, lower left, right 3, 84, 85, 100, 101

Nutone Division of Scovill Madison and Red Bank Roads, Cincinnati, Ohio 45227. 91 lower left

Red Cedar Shingle & Handsplit Shake Bureau 5510 White Building, Seattle, Washington 98101. 6 upper right, 7 upper right, 27 lower right, 29 upper right, 37, 76 lower right, 77, 78, 79, 80 upper left, 91 right center, 92. Ventilating drawings adapted w/permission 97 upper, 98

James E. Russell 1646 Vinton, Memphis, Tennessee 38105. 57

The Stanley Company Hand Tool Division, New Britain, Connecticut 06050. Drawings of rafter square use appearing in Chapter 11 redrawn with permission.

Western Wood Products Association Yeon Building Portland, Oregon 97204. 14

Ventilation drawings on pages 96 and 97 adapted from information from H.C. Products Co. Box 68, Princeville, Illinois 61159